Lecture Notes in Computer Science   5464

Commenced Publication in 1973
Founding and Former Series Editors:
Gerhard Goos, Juris Hartmanis, and Jan van Leeuwen

Editorial Board

David Hutchison
  Lancaster University, UK
Takeo Kanade
  Carnegie Mellon University, Pittsburgh, PA, USA
Josef Kittler
  University of Surrey, Guildford, UK
Jon M. Kleinberg
  Cornell University, Ithaca, NY, USA
Alfred Kobsa
  University of California, Irvine, CA, USA
Friedemann Mattern
  ETH Zurich, Switzerland
John C. Mitchell
  Stanford University, CA, USA
Moni Naor
  Weizmann Institute of Science, Rehovot, Israel
Oscar Nierstrasz
  University of Bern, Switzerland
C. Pandu Rangan
  Indian Institute of Technology, Madras, India
Bernhard Steffen
  University of Dortmund, Germany
Madhu Sudan
  Microsoft Research, Cambridge, MA, USA
Demetri Terzopoulos
  University of California, Los Angeles, CA, USA
Doug Tygar
  University of California, Berkeley, CA, USA
Gerhard Weikum
  Max-Planck Institute of Computer Science, Saarbruecken, Germany

Rui Valadas   Paulo Salvador (Eds.)

Traffic Management and Traffic Engineering for the Future Internet

First Euro-NF Workshop, FITraMEn 2008
Porto, Portugal, December 2008
Revised Selected Papers

 Springer

Volume Editors

Rui Valadas
Instituto de Telecomunicações, Instituto Superior Técnico
Av. Rovisco Pais, 1049-001 Lisboa, Portugal
E-mail: rui.valadas@ist.utl.pt

Paulo Salvador
Instituto de Telecomunicações, Universidade de Aveiro
Campus de Santiago, 3810-193 Aveiro, Portugal
E-mail: salvador@ua.pt

Library of Congress Control Number: 2009934595

CR Subject Classification (1998): C.2, D.4.4, G.2.2, H.3.4, C.2.5, C.2.6

LNCS Sublibrary: SL 5 – Computer Communication Networks
and Telecommunications

ISSN   0302-9743
ISBN-10   3-642-04575-8 Springer Berlin Heidelberg New York
ISBN-13   978-3-642-04575-2 Springer Berlin Heidelberg New York

This work is subject to copyright. All rights are reserved, whether the whole or part of the material is
concerned, specifically the rights of translation, reprinting, re-use of illustrations, recitation, broadcasting,
reproduction on microfilms or in any other way, and storage in data banks. Duplication of this publication
or parts thereof is permitted only under the provisions of the German Copyright Law of September 9, 1965,
in its current version, and permission for use must always be obtained from Springer. Violations are liable
to prosecution under the German Copyright Law.

springer.com

© Springer-Verlag Berlin Heidelberg 2009
Printed in Germany

Typesetting: Camera-ready by author, data conversion by Scientific Publishing Services, Chennai, India
Printed on acid-free paper   SPIN: 12756650   06/3180   5 4 3 2 1 0

Preface

Designing the future internet requires an in-depth consideration of the management, dimensioning and traffic control issues that will be involved in the network operations of these networks. The International Workshop on Traffic Management and Traffic Engineering of the Future Internet, FITraMEn 2008, organized within the framework of the Network of Excellence Euro-NF[1], provided an open forum to present and discuss new ideas in this area in the context of fixed, wireless and spontaneous (ad hoc and sensor) networks.

The Network of Excellence Euro-NF "Anticipating the Network of the Future - From Theory to Design" is a European project funded by the European Union within the Seventh Framework Program. The focus of Euro-NF is to develop new principles and methods to design/dimension/control/manage multi-technology architectures. The emerging networking paradigms raise new challenging scientific and technological problems embedded in complex policy, governance, and worldwide standards issues. Dealing with the diversity of these scientific and social, political and economic challenges requires the integration of a wide range of research capabilities, a role that Euro-NF aims to fulfill.

This proceedings volume contains a selection of the research contributions presented at FITraMEn 2008. The workshop was held December 11–12, 2008 in Porto, Portugal, organized by Instituto de Telecomunicações[2].

The workshop was attended by a total of 44 participants, 12 external to the Euro-NF community. The technical program included, in addition to the paper presentations, two keynote speeches, one by Nandita Dukkipati of Advanced Architecture and Research at Cisco Systems, San Jose, on "Challenges in the Traffic Management of Future Networks" and the other by Andreas Kind, from IBM Zurich Research Laboratory, on "Monitoring the Future Internet."

To participate in the workshop the authors submitted a full paper on their recent research work in the field of traffic management and traffic engineering. The 54 papers submitted to the workshop were carefully reviewed by at least 3 members of the Technical Program Committee. A total of 30 papers were selected for oral presentation at the workshop. From these, 15 revised papers incorporating the numerous referee comments and the suggestions made in the workshop were accepted for publication in this book.

The papers contained in this book provide a general view of the ongoing research on traffic management and traffic engineering in the Euro-NF Network of Excellence, and give a representative example of the problems currently

---

[1] http://euronf.enst.fr
[2] http://www.it.pt

investigated in this area, spanning topics such as bandwidth allocation and traffic control, statistical analysis, traffic engineering, and optical networks and video communications.

March 2009   Rui Valadas
                                                                 Paulo Salvador

Organization

Organizing Committee

Rui Valadas	Institute of Telecommunications, Portugal
Paulo Salvador	Institute of Telecommunications, Portugal
António Nogueira	Institute of Telecommunications, Portugal
Joel Rodrigues	Institute of Telecommunications, Portugal
Susana Sargento	Institute of Telecommunications, Portugal
Eduardo Rocha	Institute of Telecommunications, Portugal
Sandra Corújo	Institute of Telecommunications, Portugal

Referees

Alvelos, Filipe
Antunes, Nelson
Atmaca, Tulin
Borst, Sem
Brown, Patrick
Casaca, Augusto
Casetti, Claudio
Collange, Denis
Dotaro, Emmanuel
Fiedler, Markus
Garcia, Nuno
Jajszczyk, Andrzej
Johansson, Mikael

Kilpi, Jorma
Krieger, Udo
Mellia, Marco
Menth, Michael
Naldi, Maurizio
Nogueira, António
Norros, Ilkka
Oliveira, Rosário
Oriolo, Gianpaolo
Pacheco, António
Pereira, Paulo
Pioro, Michal
Roberts, James

Rodrigues, Joel
Sabella, Roberto
Salvador, Paulo
Sargento, Susana
Sidi, Moshe
de Sousa, Amaro
Tassiulas, Leandros
Valadas, Rui
Vaton, Sandrine
Virtamo, Jorma
Zúquete, André

Sponsoring Institutions

EU-funded Euro-NF network of Excellence
Institute of Telecommunications, Portugal

Table of Contents

Optical Networks and Video Communications

Models for Capacity Demand Estimation in a TV Broadcast Network with Variable Bit Rate TV Channels

Zlatka Avramova[1], Danny De Vleeschauwer[2], Kathleen Spaey[3],
Sabine Wittevrongel[1], Herwig Bruneel[1], and Chris Blondia[3]

[1] SMACS Research Group, TELIN Department, Ghent University,
Sint–Pietersnieuwstraat 41, B–9000 Ghent, Belgium
{kayzlat,sw,hb}@telin.ugent.be
[2] Bell Labs, Alcatel-Lucent Bell, Copernicuslaan 50, B–2018 Antwerp, Belgium
danny.de_vleeschauwer@alcatel-lucent.be
[3] PATS Research Group, IBBT–University of Antwerp, Middelheimlaan 1, B–2020
Antwerp, Belgium
{kathleen.spaey,chris.blondia}@ua.ac.be

Abstract. Mobile TV is growing beyond the stage of experimentation
and evaluation and is (about) to become part of our daily lives. Additionally, it is being delivered through heterogeneous networks and to a
variety of receiving devices, which implies different versions of one and
the same video content must be transported. We propose two (approximate) analytic methods for capacity demand estimation in a (mobile)
TV broadcast system. In particular, the methods estimate the required
transport capacity for a bouquet of channels offered on request and in
different versions (video formats or in different quality) over a multicast-
enabled network, encoded in non-constant bit rate targeting constant
quality. We compare a transport strategy where the different versions
(of one channel) are simulcast to a scalable video encoding (SVC) transport strategy, where all resolutions (of one channel) are embedded in one
flow. In addition, we validate the proposed analytic methods with simulations. A realistic mobile TV example is considered with two transported
resolutions of the channels: QVGA and VGA. We demonstrate that not
always capacity gain is achieved with SVC as compared to simulcast
since the former comes with some penalty rate and the gain depends on
the system parameters.

1 Introduction

The compression efficiency of the new codecs H.264/MPEG-4 AVC [1] with
respect to the codecs MPEG-2 [2] and MPEG-4 Part 2 [3] makes that the former
codecs are imposing more and more in digital (IP-based) multimedia delivery
systems. An extension to the AVC codec, called SVC (Scalable Video Coding) [4],
allows for scalable, multi-layer encoding. A base layer ensures some basic quality
of experience (e.g., targets a small-resolution device) and is complemented by one

R. Valadas and P. Salvador (Eds.): FITraMEn 2008, LNCS 5464, pp. 1–15, 2009.
© Springer-Verlag Berlin Heidelberg 2009

or several enhancement layers improving the video quality or targeting higher-resolution display screens. We consider a broadcast network where a number of channels are subject to multicast streaming to receivers of different classes of devices (resolutions) or different quality classes. The question arises which of the two following scenarios requires the least transport capacity. The first scenario simulcasts all resolutions (encoded in H.264 AVC) of all channels that are requested by at least one user. The second scenario exploits the scalability property of H.264 SVC that the bit stream of one particular resolution embeds all lower resolutions and transmits only the highest resolution that is requested by at least one user. Since all resolutions below that particular resolution piggy-back along in the stream, no other resolutions need to be transmitted in parallel in this scenario.

In [5], we demonstrated for constant bit rate (CBR) encoded video that the answer to the question formulated above is not straightforward. It is basically a trade-off between the fact that SVC encoding comes with a certain penalty rate and the fact that simulcast sends all the requested resolutions in parallel, while SVC has to transmit only one embedded stream. On the other hand, this embedded stream in the SVC case has to contain all the layers up to the highest requested resolution regardless of whether the lower ones are requested or not, while simulcast only sends the resolutions that are requested at least by one user. Whereas in our previous work [5] we considered that the channels are CBR encoded, here we describe the bit rate fluctuation of a channel stochastically by a given distribution based on real data. The extended model takes into account the fact that the required bit rates for the resolutions of a given channel are content dependent and correlated.

The rest of the paper is structured as follows. In Sect. 2, we describe the system under study. We propose two analytic methods for capacity demand estimation in Sect. 3 and 4. Our simulation approach is presented in Sect. 5. Numerical results are discussed in Sect. 6, and finally the paper is concluded in Sect. 7.

2   System under Study

We aim at developing models to estimate the capacity demand on the aggregation link in a (mobile) TV broadcast system of digital (IP-transported) channels, where a bouquet of TV channels is offered to an audience of subscribers. Aggregation network may imply also the wireless radio interface depending on the particular network design. A bouquet of K TV channels are broadcast (multicast) to the (active) subscribers (users) on request. The bit rate of a streamed channel is not constant, since constant quality is aimed for by the operator. Moreover, a system with multi-resolution receiver devices is envisioned, requiring that the content of the streamed channels is encoded in L different resolutions, either by providing separate independent versions (simulcast scenario) or by encoding them in scalable embedded/multi-layered streams (e.g., in SVC). The bouquet of offered channels is characterised by a given channel popularity distribution from which the probability ρ_k with $k \in \{1, \ldots, K\}$ that a certain channel is requested for watching is drawn. Very often, e.g. in [6], this distribution is approximated by a Zipf law [7]:

$$\rho_k = dk^{-\alpha}, \tag{1}$$

where d is a normalisation constant ensuring that the sum of all the probabilities ρ_k is 1, and α is the parameter of the Zipf distribution, ranging typically between 0.5 and 1.

We denote the *capacity demand* $\mathbf{C}$ under a simulcast scenario by $\mathbf{C}_{SIM}$, and under the SVC scenario by $\mathbf{C}_{SVC}$. Remark that these are random variables, since there is randomness in a user's behaviour: a user watches television or not, he selects a TV channel (respectively streamed video content) among K available channels, he is requesting it in a given resolution. We also assume that if the multicast channels are scalably encoded, only the layers up to and including the highest requested one are multicast, since otherwise, it would result in a waste of network resources.

Once a user is tuned in to a particular channel, and in contrast to [5], we assume that the required bit rate of this channel is fluctuating. We only consider fluctuations associated with the succession of scenes of programmes transported over that (selected) channel (because we assume that fluctuations associated with frame type are smoothened by the shaper in the encoder [8]). In the general case, we assume that there are no dependencies between the bit rates of the different channels but there is a dependence between the rates of the different resolutions of one channel. Under resolution we mean the different versions under simulcast (SIM) streaming or the different layers in scalable encoding of a channel under SVC streaming. This correlation between the different resolutions accounts for the fact that if a high bit rate is produced in one scene, this is probably the case for all versions/layers of a channel transporting the same content. With this assumption, a multicast channel's resolution generates with rate $\mathbf{r}_\ell$, where ℓ stands for the resolution and $\ell \in \{1, \ldots, L\}$. Remark that for simulcast (SIM) each resolution ℓ has its own specific bit rate $\mathbf{r}_\ell$, while for SVC to subscribe to resolution ℓ all layers up to the streamed layer ℓ need to be received, i.e., in the SVC case we denote by $\mathbf{r}_\ell$ the total bit rate of all layers 1 to ℓ. The channel's versions/layers have a joint distribution $\Pi_{\{\mathbf{r}_1,\ldots,\mathbf{r}_\ell,\ldots,\mathbf{r}_L\}}(r_1, \ldots, r_\ell, \ldots, r_L)$ with L marginal distributions $\{\pi_{\mathbf{r}_\ell}\}$ corresponding to the distribution associated to resolution ℓ from which its rate is drawn. In this work, we assume that all the channels behave statistically the same, hence have the same joint distribution (and resulting marginal distributions). In the first analytical method we will consider, we will only need the vectors of the resolutions' average rates and the covariance matrices. In the second analytical approach, a recursive method, the distribution of the rates of the resolutions is approximated by a discrete histogram in order to be able to calculate the convolutions of histograms (marginal distributions) numerically approximating probability density functions.

Furthermore, we assume that in all offered resolutions the channel popularity distribution (the Zipf distribution) is the same, i.e., a user associated to a given resolution selects which channel to watch according to the popularity law given in (1) irrespective of the resolution.

2.1   Capacity Demand under Simulcast

We define the random variables $\mathbf{n}_{k,\ell}$, where $1 \leq k \leq K, 1 \leq \ell \leq L$, as the number of users watching channel k in resolution ℓ. Because with simulcast every channel is encoded in L independent versions, and a channel is streamed in version ℓ if at least one user watches that channel in resolution (version) ℓ, we can express $\mathbf{C}_{SIM}$ in the following way:

$$\mathbf{C}_{SIM} = \sum_{k=1}^{K}\sum_{\ell=1}^{L} r_\ell 1_{\{\mathbf{n}_{k,\ell}>0\}}, \tag{2}$$

where $1_{\{m\}}$ is the indicator function of the event m, expressing that there is a contribution to $\mathbf{C}_{SIM}$ from a channel k if it is watched by at least one user in resolution ℓ.

2.2   Capacity Demand under SVC

With scalable video encoding, in order to watch a channel in resolution ℓ, all layers 1 until ℓ of that channel are needed for decoding, because they are all interrelated (in this paper we do not consider scalable encoding such as e.g., multiple description coding, MDC). Thus, layer 1 until ℓ of channel k need to be transported if there is at least one user watching channel k in resolution ℓ, and if there is no user watching channel k in a higher resolution than ℓ. Therefore, $\mathbf{C}_{SVC}$ can be expressed as follows:

$$\mathbf{C}_{SVC} = \sum_{k=1}^{K}\sum_{\ell=1}^{L} r_\ell 1_{\{\mathbf{n}_{k,L}=0,\ldots,\mathbf{n}_{k,\ell+1}=0,\mathbf{n}_{k,\ell}>0\}}. \tag{3}$$

2.3   Channel Viewing Probabilities

We assume that a user (with his receiving device) is associated only to a given resolution. The group of N users (subscribers) is divided in L fixed sets. Within a set, all clients use a terminal capable of receiving resolution ℓ, and there are N_ℓ users of type ℓ so that the sum of all N_ℓ over all L resolutions/sets is N.

The TV channels are assumed to have independent probabilities ρ_k of being watched, which are proportional to the popularity of the channels, drawn as explained above from a Zipf distribution.

We calculate the following probability generating function for the joint probability of the channels being watched by a given set of users:

$$F(z_{k,\ell}; \forall k, \ell) = E\left[\prod_{k=1}^{K}\prod_{\ell=1}^{L} z_{k,\ell}^{\mathbf{n}_{k,\ell}}\right], \tag{4}$$

where $\sum_{k=1}^{K} \mathbf{n}_{k,\ell} = \mathbf{n}_\ell \leq N_\ell$.

Inside a set of N_ℓ users associated to layer ℓ, a user has an activity grade a_ℓ, the probability of being active watching a channel, which is the average fraction of the time that a user of type ℓ watches television. We then have that the sample space corresponding to the vector of random variables $(\mathbf{n}_{1,1}, \mathbf{n}_{2,1}, \ldots, \mathbf{n}_{K,L})$ is constituted by all tuples $(n_{1,1}, n_{2,1}, \ldots, n_{K,L})$ for which $\sum_{k=1}^{K} n_{k,\ell} = n_\ell \leq N_\ell$ (note that the random variables are expressed in Bold typeface, unlike the values they assume). The probability corresponding to a realisation of the variable, expressed by the tuple $(n_{1,1}, n_{2,1}, \ldots, n_{K,L})$, is given by:

$$
\Pr \begin{bmatrix} \mathbf{n}_{1,1} = n_{1,1}, \; \mathbf{n}_{2,1} = n_{2,1}, \; \ldots, \mathbf{n}_{K,1} = n_{K,1}, \\ \cdots \\ \mathbf{n}_{1,\ell} = n_{1,\ell}, \; \mathbf{n}_{2,\ell} = n_{2,\ell}, \; \ldots, \mathbf{n}_{K,\ell} = n_{K,\ell}, \\ \cdots \\ \mathbf{n}_{1,L} = n_{1,L}, \mathbf{n}_{2,L} = n_{2,L}, \; \ldots, \mathbf{n}_{K,L} = n_{K,L} \end{bmatrix}
$$

$$
= \prod_{\ell=1}^{L} \left[\frac{N_\ell!}{(N_\ell - n_\ell)!} (1 - a_\ell)^{N_\ell - n_\ell} a_\ell^{n_\ell} \prod_{k=1}^{K} \frac{(\rho_k)^{n_{k,\ell}}}{n_{k,\ell}!} \right], \tag{5}
$$

to which corresponds the following probability generating function:

$$
F(z_{k,\ell}; \forall k, \ell) = \prod_{\ell=1}^{L} \left[1 - a_\ell + a_\ell \sum_{k=1}^{K} \rho_k z_{k,\ell} \right]^{N_\ell}. \tag{6}
$$

The channel viewing probabilities associated to other user models can be calculated by first deriving the probability generating function corresponding to the model.

3 Capacity Demand – A Gaussian Approximation Approach

We assume first that the aggregate capacity demand is a Gaussian variable, thus, provided its average and variance (standard deviation) are known, the distribution can be found. For practical reasons we express it in the form of the complementary cumulative distribution function (CCDF), also referred to as tail distribution function (TDF). This is the probability that the required bandwidth exceeds a certain value C_{avail}, i.e., $\Pr[\{\mathbf{C}_{SIM}, \mathbf{C}_{SVC}\} > C_{avail}]$.

3.1 Average of the Aggregate Flow

The average of the aggregate flow $\mathbf{C}$ from multicast channels in L resolutions is the sum of the averages of the contributions of all the resolutions in both simulcast (SIM) or scalable encoding (SVC) scenarios.

Simulcast. In the simulcast case, the mean of the aggregate traffic $\mathbf{C}_{SIM}$ is:

$$
E[\mathbf{C}_{SIM}] = \sum_{\ell=1}^{L} \sum_{k=1}^{K} E\left[\mathbf{r}_\ell \mathbf{1}_{\{\mathbf{n}_{k,\ell}>0\}}\right] = \sum_{\ell=1}^{L} \sum_{k=1}^{K} E[\mathbf{r}_\ell] E\left[\mathbf{1}_{\{\mathbf{n}_{k,\ell}>0\}}\right]
$$

$$
= \sum_{\ell=1}^{L} E[\mathbf{r}_\ell] \sum_{k=1}^{K} \Pr[\mathbf{n}_{k,\ell} > 0], \tag{7}
$$

where $E[\mathbf{r}_\ell]$ is the average rate of resolution ℓ. Here we have used the property that the average of an indicator function is the probability of its event. $\Pr[\mathbf{n}_{k,\ell} > 0]$ expresses the probability that a given channel k is requested in the given resolution ℓ by at least one user (hence the channel is multicast streamed).

SVC. In the SVC case, the mean of the aggregate traffic is:

$$
\begin{aligned}
E[\mathbf{C}_{SVC}] &= \sum_{\ell=1}^{L}\sum_{k=1}^{K} E\big[\mathbf{r}_\ell 1_{\{\mathbf{n}_{k,L}=0,\dots,\mathbf{n}_{k,\ell+1}=0,\mathbf{n}_{k,\ell}>0\}}\big] \\
&= \sum_{\ell=1}^{L}\sum_{k=1}^{K} E[\mathbf{r}_\ell]\, E\big[1_{\{\mathbf{n}_{k,L}=0,\dots,\mathbf{n}_{k,\ell+1}=0,\mathbf{n}_{k,\ell}>0\}}\big] \\
&= \sum_{\ell=1}^{L} E[\mathbf{r}_\ell] \sum_{k=1}^{K} \Pr[\mathbf{n}_{k,L}=0,\dots,\mathbf{n}_{k,\ell+1}=0,\mathbf{n}_{k,\ell}>0],
\end{aligned}
\tag{8}
$$

where $E[\mathbf{r}_\ell]$ stands for the average contribution of a scalable flow of up to and including layer ℓ.

3.2 Variance of the Aggregate Flow

In order to obtain the variance, first the second-order moments of the aggregate traffic under the SIM and SVC scenarios (respectively of $\mathbf{C}_{SIM}$ and $\mathbf{C}_{SVC}$) are derived below. With the index k for the probabilities that a channel k is watched by a given number of users $\mathbf{n}_{k,\ell}$, we make distinction between the channels, as there is correlation between the rates of the versions/layers of one and the same channel, but there is no correlation between those of different channels.

Simulcast. The second-order moment of $\mathbf{C}_{SIM}$ is expressed as follows:

$$
\begin{aligned}
E[\mathbf{C}_{SIM}^2] &= \sum_{k_1=1}^{K}\sum_{k_2=1}^{K}\sum_{\ell_1=1}^{L}\sum_{\ell_2=1}^{L} E\big[\mathbf{r}_{\ell_1}\mathbf{r}_{\ell_2} 1_{\{\mathbf{n}_{k_1,\ell_1}>0,\mathbf{n}_{k_2,\ell_2}>0\}}\big] \\
&= \sum_{k=1}^{K}\sum_{\ell=1}^{L} E[\mathbf{r}_\ell^2]\, E\big[1_{\{\mathbf{n}_{k,\ell}>0\}}\big] + \sum_{k=1}^{K}\sum_{\ell_1=1}^{L}\sum_{\substack{\ell_2=1 \\ \ell_2\neq\ell_1}}^{L} E[\mathbf{r}_{\ell_1}\mathbf{r}_{\ell_2}]\, E\big[1_{\{\mathbf{n}_{k,\ell_1}>0,\mathbf{n}_{k,\ell_2}>0\}}\big] \\
&\quad + \sum_{k_1=1}^{K}\sum_{\substack{k_2=1 \\ k_2\neq k_1}}^{K}\sum_{\ell_1=1}^{L}\sum_{\ell_2=1}^{L} E[\mathbf{r}_{\ell_1}]\,E[\mathbf{r}_{\ell_2}]\, E\big[1_{\{\mathbf{n}_{k_1,\ell_1}>0,\mathbf{n}_{k_2,\ell_2}>0\}}\big] \\
&= \sum_{k=1}^{K}\sum_{\ell=1}^{L} E[\mathbf{r}_\ell^2]\, \Pr[\mathbf{n}_{k,\ell}>0] + \sum_{k=1}^{K}\sum_{\ell_1=1}^{L}\sum_{\substack{\ell_2=1 \\ \ell_2\neq\ell_1}}^{L} E[\mathbf{r}_{\ell_1}\mathbf{r}_{\ell_2}]\, \Pr[\mathbf{n}_{k,\ell_1}>0,\mathbf{n}_{k,\ell_2}>0] \\
&\quad + \sum_{k_1=1}^{K}\sum_{\substack{k_2=1 \\ k_2\neq k_1}}^{K}\sum_{\ell_1=1}^{L}\sum_{\ell_2=1}^{L} E[\mathbf{r}_{\ell_1}]\,E[\mathbf{r}_{\ell_2}]\, \Pr[\mathbf{n}_{k_1,\ell_1}>0,\mathbf{n}_{k_2,\ell_2}>0].
\end{aligned}
\tag{9}
$$

In (9) we accounted for the fact that the probabilities of the bit rates are independent from the probabilities of requesting a channel, and also for the fact that the rates of the different resolutions of one and the same channel are correlated. The following relation holds for the covariance: $\text{Cov}[r_{\ell_1}, r_{\ell_2}] = E[r_{\ell_1} r_{\ell_2}] - E[r_{\ell_1}] E[r_{\ell_2}]$. The terms $E[r_{\ell_1} r_{\ell_2}]$ in (9) are thus directly related to the covariance terms $\text{Cov}[r_{\ell_1}, r_{\ell_2}]$ which compose a covariance matrix, the main diagonal of which is formed by the variance terms $\text{Var}[r_\ell]$ related to $E[r_\ell^2]$.

The square of the average of $\mathbf{C}_{SIM}$ can be also expressed as three types of terms corresponding to the three types of terms in (9). The variance of the aggregate traffic $\mathbf{C}_{SIM}$ is $\text{Var}[\mathbf{C}_{SIM}] = E[\mathbf{C}_{SIM}^2] - E[\mathbf{C}_{SIM}]^2$. Under simulcast, the variance of $\mathbf{C}_{SIM}$ has the following form:

$$
\begin{aligned}
\text{Var}[\mathbf{C}_{SIM}] = {} & \sum_{k=1}^{K} \sum_{\ell=1}^{L} \Pr[\mathbf{n}_{k,\ell} > 0] \left(E[r_\ell^2] - E[r_\ell]^2 \Pr[\mathbf{n}_{k,\ell} > 0] \right) \\
& + \sum_{k=1}^{K} \sum_{\ell_1=1}^{L} \sum_{\substack{\ell_2=1 \\ \ell_2 \neq \ell_1}}^{L} \Big(E[r_{\ell_1} r_{\ell_2}] \Pr[\mathbf{n}_{k,\ell_1} > 0, \mathbf{n}_{k,\ell_2} > 0] \\
& \qquad - E[r_{\ell_1}] E[r_{\ell_2}] \Pr[\mathbf{n}_{k,\ell_1} > 0] \Pr[\mathbf{n}_{k,\ell_2} > 0] \Big) \\
& + \sum_{k_1=1}^{K} \sum_{\substack{k_2=1 \\ k_2 \neq k_1}}^{K} \sum_{\ell_1=1}^{L} \sum_{\ell_2=1}^{L} E[r_{\ell_1}] E[r_{\ell_2}] \Big(\Pr[\mathbf{n}_{k_1,\ell_1} > 0, \mathbf{n}_{k_2,\ell_2} > 0] \\
& \qquad - \Pr[\mathbf{n}_{k_1,\ell_1} > 0] \Pr[\mathbf{n}_{k_2,\ell_2} > 0] \Big).
\end{aligned}
\tag{10}
$$

If we denote by $\mathbf{R}_{k\ell}$ the variable $r_\ell \mathbf{1}_{\{\mathbf{n}_{k,\ell}>0\}}$, we can rewrite more concisely and meaningfully the equation above. The terms in the first double sum in the equation above correspond to the sum of the variances of the contribution of resolution ℓ of channel k to the aggregate flow, $\text{Var}[\mathbf{R}_{k\ell}]$. Similarly, the second type of terms correspond to the covariance between two resolutions of a channel, and the third type of terms correspond to the contribution of the covariance between channels. The equation above can be replaced by the more elegant one given below:

$$
\begin{aligned}
\text{Var}[\mathbf{C}_{SIM}] = {} & \sum_{k=1}^{K} \sum_{\ell=1}^{L} \text{Var}[\mathbf{R}_{k\ell}] + \sum_{k=1}^{K} \sum_{\ell_1=1}^{L} \sum_{\substack{\ell_2=1 \\ \ell_2 \neq \ell_1}}^{L} \text{Cov}[\mathbf{R}_{k\ell_1}, \mathbf{R}_{k\ell_2}] \\
& + \sum_{k_1=1}^{K} \sum_{\substack{k_2=1 \\ k_2 \neq k_1}}^{K} \sum_{\ell_1=1}^{L} \sum_{\ell_2=1}^{L} \text{Cov}[\mathbf{R}_{k_1\ell_1}, \mathbf{R}_{k_2\ell_2}].
\end{aligned}
\tag{11}
$$

SVC. The formula for the variance of $\mathbf{C}_{SVC}$ is the same as (10) but with the following substitutions:

$\Pr[\mathbf{n}_{k,\ell} > 0]$ replaced by

$$\Pr[\mathbf{n}_{k,L} = 0, \ldots, \mathbf{n}_{k,\ell+1} = 0, \mathbf{n}_{k,\ell} > 0] \, ; \tag{12}$$

$\Pr[\mathbf{n}_{k,\ell_1} > 0, \mathbf{n}_{k,\ell_2} > 0]$ (where $\ell_1 \neq \ell_2$) replaced by 0; $\qquad$ (13)

$\Pr[\mathbf{n}_{k_1,\ell_1} > 0, \mathbf{n}_{k_2,\ell_2} > 0]$ (where $k_1 \neq k_2$) replaced by

$$\Pr[\mathbf{n}_{k_1,L} = 0, \ldots, \mathbf{n}_{k_1,\ell_1+1} = 0, \mathbf{n}_{k_1,\ell_1} > 0,$$
$$\mathbf{n}_{k_2,L} = 0, \ldots, \mathbf{n}_{k_2,\ell_2+1} = 0, \mathbf{n}_{k_2,\ell_2} > 0] \, . \tag{14}$$

3.3   Calculation of the Probabilities

In Sect. 2.3, the channel viewing probabilities were defined. If the probability generating function $F(z_{k,\ell}; \forall k, \ell)$ associated to a certain user behaviour model is known, then the probability that some of the $\mathbf{n}_{k,\ell}$ are equal to 0 is obtained as follows: those $z_{k,\ell}$ corresponding to $\mathbf{n}_{k,\ell} = 0$ are set to 0, the rest of the arguments $z_{k,\ell}$ are set to 1. Let $\mathbf{n}_{k,\ell} = 0$ be the event A; let $\mathbf{n}_{k,L} = 0, \ldots, \mathbf{n}_{k,\ell+1} = 0$ be the event B and $\mathbf{n}_{k,L} = 0, \ldots, \mathbf{n}_{k,\ell+1} = 0, \mathbf{n}_{k,\ell} = 0$ be the event C. Additionally, we put a subscript 1 or respectively 2 if ℓ (and k) has a subscript 1 or respectively 2. The following relations are taken into account to derive the probabilities appearing in equations (7)–(14):

$$\Pr[\bar{A}] = 1 - \Pr[A] \, ,$$
$$\Pr[\bar{A}_1 \cap \bar{A}_2] = 1 - \Pr[A_1] - \Pr[A_2] + \Pr[A_1 \cap A_2] \, ,$$
$$\Pr[\bar{A} \cap B] = \Pr[B] - \Pr[C] \, ,$$
$$\Pr[\bar{A}_1 \cap \bar{A}_2 \cap B_1 \cap B_2]$$
$$= \Pr[B_1 \cap B_2] - \Pr[C_1 \cap B_2] - \Pr[B_1 \cap C_2] + \Pr[C_1 \cap C_2] \, . \tag{15}$$

4   Capacity Demand – A Recursive Analytical Approach

Enumerating every possible outcome of the random variable of the capacity demand and calculating its associated probability would lead to a combinatorial explosion. Therefore, we use a recursive approach in this section (an extension of the "divide and conquer approach" in [5]) in order to calculate the CCDF (TDF) of the capacity demand. Since the bit rate of a resolution of a channel is no longer deterministic as in [5], but described by a probability density function (pdf), we will need to approximate this pdf by a discretely binned histogram, which has as consequence that this approach is, in contrast to [5], no longer exact. We define for every $\ell \in \{1, \ldots, L\}$ the random variables $\mathbf{w}_\ell$ and $\mathbf{c}_\ell$, where $\mathbf{w}_\ell$ represents the number of users that watch a channel in resolution ℓ; in the simulcast case $\mathbf{c}_\ell$ denotes the number of channels that need to be provided in version ℓ, while in the SVC case $\mathbf{c}_\ell$ denotes the number of channels for which layers 1 until ℓ need to be transported, but no layers higher than ℓ. Let $\mathcal{C}$, respectively $\mathcal{W}$, denote the set of all possible values the vector of random variables $(\mathbf{c}_1, \ldots, \mathbf{c}_L)$, respectively $(\mathbf{w}_1, \ldots, \mathbf{w}_L)$, can take. In [5], an exact recursive method was presented to calculate the probabilities $\Pr[(\mathbf{c}_1, \ldots, \mathbf{c}_L) = (c_1, \ldots, c_L)|(\mathbf{w}_1, \ldots, \mathbf{w}_L) = (w_1, \ldots, w_L)]$ for all $(c_1, \ldots, c_L) \in \mathcal{C}$ and all $(w_1, \ldots, w_L) \in \mathcal{W}$.

Then by the theorem on total probability, we obtain for all $(c_1, \ldots, c_L) \in \mathcal{C}$:

$$\Pr[(\mathbf{c}_1, \ldots, \mathbf{c}_L) = (c_1, \ldots, c_L)] =$$
$$\sum_{(w_1, \ldots, w_L) \in \mathcal{W}} \Pr[(\mathbf{c}_1, \ldots, \mathbf{c}_L) = (c_1, \ldots, c_L)|(\mathbf{w}_1, \ldots, \mathbf{w}_L) = (w_1, \ldots, w_L)]$$
$$\cdot \Pr[(\mathbf{w}_1, \ldots, \mathbf{w}_L) = (w_1, \ldots, w_L)]. \tag{16}$$

Note that $\mathcal{W}$ and $\Pr[(\mathbf{w}_1, \ldots, \mathbf{w}_L) = (w_1, \ldots, w_L)]$ depend on the considered user model.

With all $(c_1, \ldots, c_L) \in \mathcal{C}$, there corresponds a probability distribution of the bandwidth required for providing the channels in the desired resolutions:

$$\{\pi_{\mathbf{C}(c_1, \ldots, c_L)}\} = (\bigotimes_{i=1}^{c_1} \{\pi_{\mathbf{r}_1}\}) \otimes (\bigotimes_{i=1}^{c_2} \{\pi_{\mathbf{r}_2}\}) \otimes \cdots \otimes (\bigotimes_{i=1}^{c_L} \{\pi_{\mathbf{r}_L}\}), \tag{17}$$

where $\otimes$ denotes the convolution operation. If we then weigh these distributions with the probabilities calculated in (16), the overall probability distribution of the required capacity is obtained:

$$\{\pi_{\mathbf{C}}\} = \sum_{(c_1, \ldots, c_L) \in \mathcal{C}} \{\pi_{\mathbf{C}(c_1, \ldots, c_L)}\} \Pr[(\mathbf{c}_1, \ldots, \mathbf{c}_L) = (c_1, \ldots, c_L)]. \tag{18}$$

Note that convolution of two probability distributions containing both n outcomes gives a new probability distribution with n^2 outcomes (less in case some combinations of outcomes result in the same outcomes). This means that the probability distributions $\{\pi_{\mathbf{C}(c_1, \ldots, c_L)}\}$ in (17) have approximately $n^{c_1 + \cdots + c_L}$ outcomes, assuming each $\{\pi_{\mathbf{r}_\ell}\}$ contains n outcomes. So, it is clearly not realistic to calculate the exact distribution $\{\pi_{\mathbf{C}}\}$ as described above. Therefore, in practice we do not work in (17) with the exact probability distributions $\{\pi_{\mathbf{r}_\ell}\}$, but rather with histograms derived for them. For the numerical examples presented further on in this paper, we first calculated the maximal possible capacity value $maxC$ corresponding to the considered scenario. We then divided the capacity range $[0, maxC]$ in bins of a certain width. The width of the bins, which we denote by $unit$, was chosen in such a way that $[0, maxC]$ is divided in a predefined number of intervals. We then constructed histograms for the $\{\pi_{\mathbf{r}_\ell}\}$'s by dividing their range in intervals of type $[m \cdot unit, (m+1) \cdot unit[$, $m = 0, 1, \ldots$ After applying first formula (17) on these histograms, and then formula (18), we end up with a histogram $\{\pi_{\mathbf{C}}^*\}$ for $\{\pi_{\mathbf{C}}\}$. Of course, the use of histograms will cause the resulting $\{\pi_{\mathbf{C}}^*\}$ to be an approximation of the exact $\{\pi_{\mathbf{C}}\}$. The smaller $unit$ is chosen, the better the approximation of $\{\pi_{\mathbf{C}}\}$ will be, since the histogram of $\{\pi_{\mathbf{C}}^*\}$ will consist of more bins.

Notice that in this section we did not take into account the correlation between the bit rates associated with the different resolutions of the same channel. In the SVC case it does not matter that all layers are correlated as all layers associated with the highest resolution that is demanded need to be transported. Only the fluctuations of the totality of all transported layers matter. For the simulcast transport mode, the correlation between bit rates of various resolutions does matter, and as such this recursive method neglects it. Due to the nature of the recursive method, it is very hard to take this correlation into account.

5   Simulation Approaches

We wrote an event-driven C-based simulator, which can simulate both simulcast and SVC streaming transport modes. This simulator generates a number of realisations of the variable *capacity demand* $\mathbf{C}$ ($\mathbf{C}_{SIM}$ under simulcast and $\mathbf{C}_{SVC}$ under SVC). For every realisation, a user is either tuned into a given channel k and resolution ℓ or is inactive (governed by the user model and activity grade). According to the selected transport mode (simulcast (SIM) or SVC), the activity grade of the users, and the popularity distribution of the multimedia content (i.e., the Zipf parameter α), we measure the TDF of the variable $\mathbf{C}$, over a sufficiently large number of realisations, depending on the required accuracy. The random number generating function interpreting the activity grade of users follows a uniform distribution.

A multivariate Gaussian variable for the rates of a channel's resolutions is generated at every simulation step, drawn from a distribution defined by the input average vector and covariance matrix. In this paper, we refer to this simulator as *Gaussian simulator*.

We constructed another (C-based) simulator too. The difference with the previous one is that the channel's resolutions rates are drawn randomly from a list of the average rates of real movies. We refer here to this simulator as *histogram simulator*. Were this list of movies' average rates long enough (sorted and possibly with assigned probabilities to every rate realisation or in a histogram form), the rates' probability distribution would have been reasonably well approximated.

6   Results

In this section we will estimate the capacity demand in a realistic mobile TV example by the four approaches described above (two analytical and two simulation ones). In this way we will validate the different approaches and will point out their advantages and their drawbacks.

6.1   Mobile TV Example Settings

For a mobile TV case, we take as guidelines the mobile TV example proposed in the DVB standardisation body [9]. All K channels are encoded in the same number of resolutions L, every resolution corresponding to a certain spatial/temporal format or quality level which is the same (or varies slightly) for all the channels from the bouquet of channels. In the mobile TV example, a channel is encoded in two spatial resolutions (with possible temporal scaling too) with a QVGA base resolution (version, layer) with bit rate in the range 250–400 kbps and a VGA resolution of 750–1500 kbps. In a simulcast scenario, two versions of a channel are encoded and transmitted, while in an SVC scenario, the VGA resolution is encoded from which either a QVGA resolution is streamed or if the VGA resolution is requested, the spatial enhancement layer is transmitted too.

We start from experimental data for 20 movies encoded in two resolutions (QVGA and VGA), both single-layered in AVC mode and multi-layered in SVC mode (encoded in baseline profile and in SVC layered mode). The two resolutions are: base resolution QVGA 320x240@15Hz for some of the movies or 12.5Hz for the other, and its spatial and temporarily scaled resolution VGA 640x480@30Hz for some of the movies or respectively 25Hz for the other. The rates are taken such that delivering a resolution in either of the transport modes (simulcast or SVC) provides the same video quality (Peak Signal-to-Noise Ratio, PSNR). Moreover, there are two encoding experiments and hence sets of input data, we refer to as SET1 and SET2. SET1 targets a low picture quality (i.e., a PSNR of approximately 32 dB), while SET2 targets a higher quality (approximately a PSNR of 34 dB) for both resolutions. Normally, the base layer of an SVC encoded video is encoded in AVC; however in the experimental data we use, the QVGA resolution in SVC has some 8% higher bit rate (for the same quality) as compared to the lowest resolution of simulcast. An SVC VGA resolution of the channel has some 21% higher bit rate than the corresponding AVC VGA version, which we refer to as "penalty rate of SVC". Note that the values specified above (8% and 21%) are the averages over all the movies. The penalty values of each movie differ.

The vectors of the resolutions' averages and covariance matrices for the two experimental sets and the two H.264 coding modes (AVC and SVC) are given in Table 1.

We set the network parameters as follows. Two cases are considered: with $K = 20$ and $K = 50$ channels. The case with a bouquet of 20 channels is representative of the currently deployed mobile TV trials and commercial deployments (as e.g., in [10]); the case of 50 channels is representative of enhanced services

Table 1. Vectors of the average bit rates and covariance matrices

Data	AVC (simulcast)		SVC	
SET1	QVGA	VGA	QVGA	VGA
Average	259	934	277	1110
Covariance	14423	49721	16274	65208
matrix	49721	176543	65208	261328
Data	AVC (simulcast)		SVC	
SET2	QVGA	VGA	QVGA	VGA
Average	436	1559	474	1899
Covariance	45924	154047	54040	216475
matrix	154047	528443	216475	867215

in future. In general, the known mobile (broadcast) TV technologies, to name but a few – DVB-H/SH (Digital Video Broadcasting-Handheld/Satellite services to Handhelds), MBMS (Multimedia Broadcast Multicast Service) (e.g., TDtv), MediaFLO, T/S-DMB (Terrestrial/Satellite-Digital Multimedia Broadcasting), etc., have the potential to deliver approximately up to 50 channels. For the parameter of the Zipf distribution, we choose $\alpha = 0.6$ as in our previous work [5]. We set the number of users N_ℓ to 100 for every of the two resolutions (QVGA and VGA) and their activity grade a_ℓ to 0.6.

6.2   Validation of the Proposed Methods

The TDFs of $\mathbf{C}_{SIM}$ and $\mathbf{C}_{SVC}$ by the two analytical and two simulation methods are displayed in Figure 1 for data SET1 and in Figure 2 for data SET2. In both figures, the upper graphs display the capacity demand TDF by the analytical Gaussian approximation approach and the TDFs obtained by the two simulation approaches (with the Gaussian simulator and the histogram simulator); in the bottom graphs, a comparison is made of the two analytical approaches (Gaussian and recursive).

Estimation of Capacity Demand. A network (broadcast) operator wants to dimension its network for a given probability of unavailability (the probability that the required network resources exceed the available bandwidth), i.e., $\Pr[\{\mathbf{C}_{SIM}, \mathbf{C}_{SVC}\} > C_{avail}]$. We call this probability $P_{unavail}$. In Table 2 we summarise the required bandwidth as estimated by the four approaches for $P_{unavail} = 10^{-4}$. For the case with 20 channels, both for SET1 and SET2, an SVC transport strategy outperforms the simulcast one: under SET1, the required capacity demand with simulcast is 31.1 Mbps while with SVC it is 29.7 Mbps; under SET2, the required capacity demand with simulcast is approximately 52.5 Mbps while with SVC it is 51.7 Mbps (these approximate figures are taken from

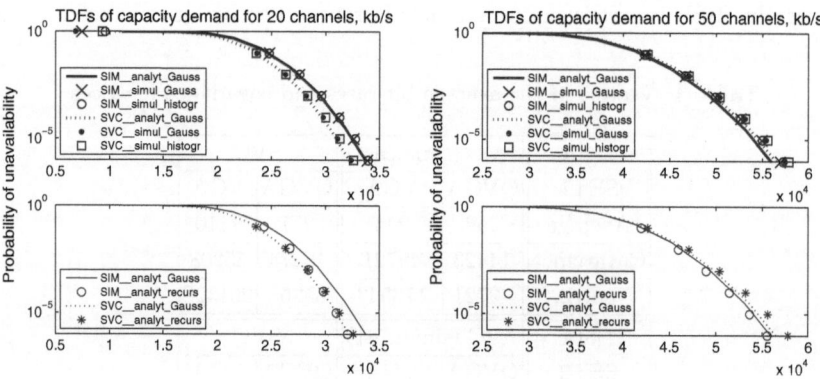

Fig. 1. Comparison of TDFs of capacity demand $\mathbf{C}_{SIM}$ and $\mathbf{C}_{SVC}$ for data SET1 by the four methods: upper graphs – by analytical Gaussian approach and the two simulation approaches; bottom graph – comparison of the two analytical approaches

Table 2. Numerical comparison of the capacity demand by the four methods at $P_{unavail} = 10^{-4}$ for the cases of Figure 1 and Figure 2

Data SET1				
Capacity demand [kbps]	$K = 20$		$K = 50$	
	simulcast	SVC	simulcast	SVC
Analytical Gaussian	31 129	29 709	51 965	52 472
Analytical recursive	29 820	30 004	51 428	53 172
Relative error	4.20%	0.99%	1.03%	1.33%
Gaussian simulator	31 206	29 778	52 617	53 111
Relative error	0.25%	0.23%	1.25%	1.22%
Histogram simulator	31 316	30 009	52 609	53 144
Relative error	0.60%	1.01%	1.24%	1.28%
Data SET2				
Capacity demand [kbps]	$K = 20$		$K = 50$	
	simulcast	SVC	simulcast	SVC
Analytical Gaussian	52 548	51 661	87 361	90 687
Analytical recursive	50 189	52 308	86 322	92 062
Relative error	4.49%	1.25%	1.19%	1.52%
Gaussian simulator	52 753	51 833	88 571	91 869
Relative error	0.39%	0.33%	1.38%	1.30%
Histogram simulator	52 883	52 309	88 531	92 051
Relative error	0.64%	1.25%	1.34%	1.50%

the analytical Gaussian approach, which accords very well with the two simulation approaches and a little less with the recursive approach, at least in the simulcast case). However, in the case of 50 channels with the same network settings, simulcast is more efficient than SVC: under SET1, the required capacity demand with simulcast is approximately 52.0 Mbps while with SVC it is 52.5 Mbps; under SET2, the required capacity demand with simulcast is approximately 87.4 Mbps while with SVC it is 90.7 Mbps.

This shows that it is not straightforward whether an SVC transport scheme will be beneficial for saving on network resources, but that this depends on the network environment (parameters). SVC becomes the more efficient transport mode when all resolutions of all channels are actively requested (for which the prerequisites are high activity grade of users, small user population, small bouquet of channels). If the number of resolutions in the broadcast system increases (and the bit rate penalty under SVC encoding remains small), the SVC mode is expected to become more effective compared to simulcasting all resolutions.

Comparison of the Four Methods. The four methods to estimate the capacity demand on an aggregation link transporting K multicast channels in L

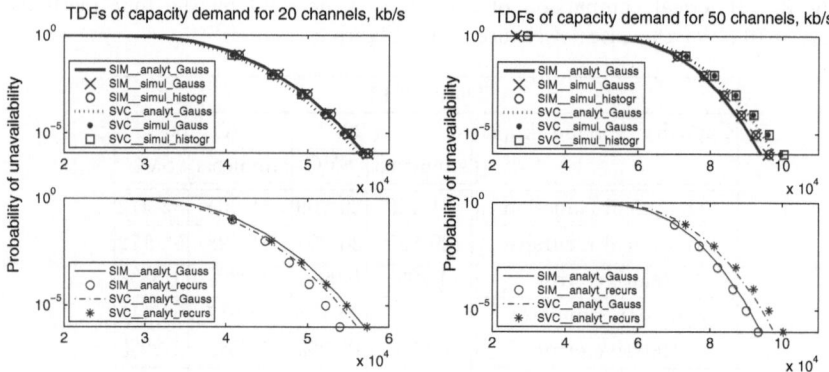

Fig. 2. Comparison of TDFs of capacity demand $\mathbf{C}_{SIM}$ and $\mathbf{C}_{SVC}$ for data SET2 by the four methods: upper graphs – by analytical Gaussian approach and the two simulation approaches; bottom graphs – comparison of the two analytical approaches

different resolutions, are consistent in the resulting TDFs of required bandwidth. However, the recursive method deviates most from the three other methods in the calculation of $\mathbf{C}_{SIM}$. The reason for this is a drawback of the approach, not allowing to account for the correlation between a channel's resolutions rates. This has no impact on the results for $\mathbf{C}_{SVC}$ since its central moments do not depend on the off-diagonal elements of the covariance matrix. This explains why the TDFs for $\mathbf{C}_{SIM}$ are qualitatively different by the two analytical methods in the lower left graphs in Figure 1 and in Figure 2. The recursive method calculates the distribution of the number of channels to be provided exactly but it needs to approximate the joint distribution of the bit rate of both resolutions by two marginal histograms. This binning and ignoring the correlations introduces an error.

The TDFs by the two simulation approaches are in a very good correspondence. Hence, it is reasonable to assume that the joint probability distribution of the bit rates of the resolutions can be approximated by a multivariate Gaussian one (under certain conditions). They show also a good match with calculated TDFs by the analytical Gaussian approach for the case of 20 channels (see the upper left graphs in Figure 1 and in Figure 2). Unfortunately, for the case of 50 channels, the analytical Gaussian method is not very exact (relative error at $P_{unavail} = 10^{-4}$ up to 1.5%). Normally however, for large network scenarios (and if the ratio N/K is large), the Gaussian analytical approach shows a good correspondence with results by the other approaches. Moreover, this method yields results in the fastest way as compared to the three other methods, it allows to calculate other user models (e.g., the ones in [5]), and it is robust to calculate large network scenarios. The drawback of the simulation methods is that the higher the required accuracy, the longer the simulation run must be.

7 Conclusion

We presented and compared four approaches to estimate the capacity demand on the aggregation link in a broadcast network with K streamed channels in L different resolutions. A realistic mobile TV scenario was calculated and simulated with two data sets (corresponding to different video quality), and the required bandwidth was predicted both in a simulcast and in an SVC transport mode. We demonstrate that not always SVC outperforms simulcast in terms of resource efficiency and that SVC becomes more beneficial with larger number of channel resolutions, smaller bit rate penalty, a more active audience of subscribers, and a small bouquet of channels.

The Gaussian approximation analytical method is the fastest one but results start to deviate with a small N/K ratio. The recursive analytical method does not account for the correlation between the bit rates of the different resolutions (which is a drawback in a simulcast scenario) and also its accuracy depends on the coarseness of the binning process. The simulation methods can be considered as most accurate of the presented approaches but their accuracy depends on the length of the simulation runs.

References

1. ITU-T Recommendation H.264 and ISO/IEC 14496-10 (MPEG-4 AVC): Advanced Video Coding for Generic Audiovisual Services. Vers.3 (March 2005)
2. ITU-T Recommendation H.262 and ISO/IEC 13818-2 (MPEG-2 Video): Generic Coding of Moving Pictures and Associated Audio Information - Part 2: Video (November 1994)
3. ITU-T Recommendation H.264 and ISO/IEC 14496-10 (MPEG-4 Visual): AVC Coding of audio-visual objects-Part 2: Visual. Vers.3 (May 2004)
4. ITU-T Recommendation H.264 and ISO/IEC 14496-10 (MPEG-4 AVC): Advanced Video Coding for Generic Audiovisual Services. Vers.8 (November 2007)
5. Avramova, Z., De Vleeschauwer, D., Spaey, K., Wittevrongel, S., Bruneel, H., Blondia, C.: Comparison of Simulcast and Scalable Video Coding in Terms of the Required Capacity in an IPTV Network. In: Proceedings of the 16th International Packet Video Workshop (PV 2007), Lausanne, Switzerland (November 2007)
6. Sinha, N., Oz, R., Vasudevan, S.: The Statistics of Switched Broadcast. In: Proceedings of Conference on Emerging Technologies (SCTE), Tampa, USA (2005)
7. Zipf, J.: Selective Studies and the Principle of Relative Frequency in Language (1932)
8. Avramova, Z., De Vleeschauwer, D., Laevens, K., Wittevrongel, S., Bruneel, H.: Modelling H.264/AVC VBR Video Traffic: Comparison of a Markov and a Self-Similar Source Model. Telecommunication Systems Journal 39, 145–156 (2008)
9. DVB-CM-AVC Document CM-AVC0182: Proposal to DVB for study work into Scalable Video Coding for IPTV (November 2007)
10. http://www.dvb-h-online.org/services.htm

A Distributed Scheme for Value-Based Bandwidth Reconfiguration

Åke Arvidsson[1], Johannes Göbel[2], Anthony Krzesinski[3,*], and Peter Taylor[4,**]

[1] Ericsson AB, Packet Technologies Research, SE-164 80 Stockholm, Sweden
Ake.Arvidsson@ericsson.com
[2] Department of Informatics, University of Hamburg, 22527 Hamburg, Germany
goebel@informatik.uni-hamburg.de
[3] Department of Mathematical Sciences, University of Stellenbosch,
7600 Stellenbosch, South Africa
aek1@cs.sun.ac.za
[4] Department of Mathematics and Statistics, University of Melbourne,
3010 Parkville, Australia
p.taylor@ms.unimelb.edu.au

Abstract. This paper presents a scheme for reallocating bandwidth in path-oriented transport networks. At random time points, bandwidth is allocated to those paths that (possibly temporarily) value it most highly. The scheme acts according to local rules and without centralised control. The proposed scheme is thus distributed and scalable. Previous studies have shown that bandwidth reallocation, together with the provision of appropriate amounts of spare capacity on certain links, can be used to rapidly deploy and capacitate equivalent recovery routes in the event of network equipment failure. The purpose of this study is to determine if the same reallocation mechanism can also deal effectively with repeated, small scale random traffic fluctuations and with time varying traffics. We present a simulation model of a 30-node 46-link network which we use to evaluate the efficacy of the bandwidth reallocation scheme. The simulation study shows that bandwidth reconfiguration can substantially reduce the connection loss probabilities.

Keywords: bandwidth prices; bandwidth reconfiguration; distributed control; network planning and optimisation; scalability.

1 Introduction

In this paper we present a distributed scheme for bandwidth reallocation that can be used in the context of a path-oriented network in which relatively long-lived paths are used to provision resources for connections or flows, whose average

---

* Supported by the South African National Research Foundation, Nokia-Siemens Networks and Telkom SA Limited.

** Supported by the Australian Research Council Centre of Excellence for the Mathematics and Statistics of Complex Systems.

R. Valadas and P. Salvador (Eds.): FITraMEn 2008, LNCS 5464, pp. 16–35, 2009.
© Springer-Verlag Berlin Heidelberg 2009

holding times are much less than the path lifetimes. Possible environments in which such a model could be useful are Multi-Protocol Label Switching (MPLS) networks where traffic engineering tunnels set up by RSVP-TE signalling act as the long-lived paths, or Asynchronous Transfer Mode (ATM) networks in which Virtual Paths (VPs) act as the long-lived paths. Other examples include the Synchronous Digital Hierarchy (SDH) or the Synchronous Optical Network (SONET).

We shall use terminology that does not suggest any of the specific environments mentioned above. Thus, we call a long-lived path a *route*. Routes traverse one or more physical links. In the present paper we restrict ourselves to connection-oriented traffic with admission controls. Hence we use the terms *connections* and *lost connections*. We shall assume that each connection in service on a route requires one unit of bandwidth on each link of the route. The number of connections that can be simultaneously carried on a route depends on the amount of bandwidth allocated to the route. However, at any point in time, it is possible that the connections in service on a route are using only a small proportion of the bandwidth allocated to that route, while the bandwidth on another route is heavily utilised. In such a case, it makes sense to transfer bandwidth from the first route to the second one, if that is possible.

We propose a scheme in which each route places a value on bandwidth, dependent on its current bandwidth and its current occupancy. Bandwidth is transferred from routes that place a low value on bandwidth to routes that place a high value on bandwidth. The scheme operates according to simple deterministic rules, using a knowledge of a route's current occupancy to perform admission control and to calculate the value of an extra unit of bandwidth and also to calculate the value that the route would lose should it give up a unit of bandwidth. The scheme then uses these values to determine whether the route should acquire or release a unit of bandwidth or do neither.

We shall view routes as being of two types: *direct* routes, which traverse just a single physical link, and *transit* routes, which traverse more than one physical link. The direct routes on the links of a transit route are referred to as its *constituent* direct routes. We assume that there is a direct route using each physical link in the network. If this is not the case, the scheme can be modified so that it is implementable, but at the expense of extra complexity.

Bandwidth values are communicated via a signalling mechanism. Signals or control packets are sent at random intervals along each transit route, recording the bandwidth valuations of the constituent direct routes. If the sum of the values of an extra unit of bandwidth on each of the constituent direct routes is greater than the cost to the transit route of losing a unit of bandwidth, then the transit route gives up a unit of bandwidth, which is taken up by each of the direct routes. Alternatively, if the sum of the costs of losing a unit of bandwidth on each of the constituent direct routes is less than the value to the transit route of an extra unit of bandwidth, then each of the direct routes gives up a unit of bandwidth, which is taken up by the transit route.

The defining feature of the scheme is that bandwidth reallocation takes place between transit routes and their constituent direct routes. Such reallocations depend on the local resource demands and bandwidth valuations. In this way the scheme is autonomous, acts without centralized control from a system coordinator and behaves entirely according to local rules. The scheme is thus distributed and scalable.

It was shown in [1] that bandwidth reallocation, together with the provisioning of appropriate amounts of spare capacity on certain links [2], can be used to rapidly deploy and capacitate equivalent recovery routes in the event of network equipment failure. The purpose of this paper is to determine if the same reallocation mechanism can also deal effectively with repeated, small scale random traffic fluctuations and with time varying traffics.

The rest of this paper is organised as follows. In Sect. 2 we briefly discuss a method for calculating bandwidth values. The bandwidth reallocation scheme is presented in Sect. 3. Results from numerical experiments to test the efficacy of the reallocation scheme are given in Sect. 4, and our conclusions are presented in Sect. 5.

2   The Value of Bandwidth

In Sect. 1 we described a situation where it is desirable to place a value on the amount of bandwidth that is allocated to a route. The question arises as to how this value should be calculated.

Much work has been done in the area of bandwidth valuation. For example Lanning *et al.* [3] studied the prices that should be charged to customers in a dynamic loss system: their perspective was that of an Internet billing system where arrival rates were user-cost dependent. Fulp and Reeves [4] concentrated on multi-market scenarios where the price of bandwidth was determined on the basis of current and future usage. In the context of rate control, Kelly *et al.* [5] and Low and Lapsley [6], proposed distributed models that optimise different types of aggregate utility as seen by sources. In these models, the price of bandwidth on a particular link was identified with the value of dual variables in a Lagrangian formulation.

We use an alternative approach to the pricing of bandwidth which was presented in [7]. For an Erlang loss system, this method computed the expected lost revenue in some interval $(0, \tau]$ due to connections being lost, conditional on the system starting in a given state. The expected lost revenue was used to compute both a buying price and selling price for bandwidth, relying on knowledge of the state at time zero.

Consider a single service, such as voice, which consumes one bandwidth unit per connection. Consider a planning horizon of τ time units. We can regard the value of U extra units of bandwidth as the difference in the total expected lost revenue over the time interval $[0, \tau]$ if the route were to increase its bandwidth by U units at time zero and if it stayed the same. Conversely, we can calculate the cost of losing U units of bandwidth as the difference in the total expected lost revenue over the time interval $[0, \tau]$ if the route were to decrease its bandwidth by U units compared with leaving it the same.

For a route with bandwidth C and whose initial state is c, let $R_{c,C}(\tau)$ denote the expected revenue lost in the interval $[0, \tau]$, given that there are c connections in progress on the route at time 0. Then the value $B_{c,C}(\tau, U)$ of U units of extra bandwidth (the buying price) and the cost $S_{c,C}(\tau, U)$ of losing U units of bandwidth (the selling price) are given by

$$B_{c,C}(\tau, U) = R_{c,C}(\tau) - R_{c,C+U}(\tau)$$

and

$$
S_{c,C}(\tau, U)
= \begin{cases} R_{c,C-U}(\tau) - R_{c,C}(\tau), & c \leq C - U \\ R_{C-U,C-U}(\tau) - R_{c,C}(\tau), & C - U < c \leq C. \end{cases}
\tag{1}
$$

The second equation in (1) was not explicitly given in [7] for the case $C - U < c \leq C$. In this case, if the route were to give up U units of bandwidth, it would also have to eject one or more of its current connections. The issue then arises as to whether an extra "penalty" value should be added to reflect the negative consequences of such a decision. The right hand side of (1) reflects the situation in which no such penalty is added. The opposite extreme would be to incorporate an "infinite" penalty, which would have the effect of precluding any bandwidth reallocation that requires the ejection of connections. For the rest of this paper, we adopt the latter course. That is, we do not allow bandwidth reallocation away from a route with more than $C - U$ connections.

Under the assumption that the system has a Poisson arrival process and exponential service times, it was shown in [7] that $R_{c,C}(\tau)$ is given by the inverse of the Laplace transform

$$\tilde{R}_{c,C}(s) = \frac{\theta \, \lambda/s \, P_c(s/\lambda)}{(s + C\mu)P_C\,(s/\lambda) - C\mu P_{C-1}\,(s/\lambda)}$$

where λ and μ are the parameters of the arrival and service processes respectively and θ is the expected revenue earned per connection,

$$P_c(s/\lambda) = (-\mu/\lambda)^c \, \Gamma_c^{(\lambda/\mu)}\,(-s/\mu)$$

and

$$\Gamma_c^{(\lambda/\mu)}\,(-s/\mu) = \sum_{k=0}^{c} \binom{c}{k}\binom{-s/\mu}{k}\left(\frac{-\lambda}{\mu}\right)^{c-k} k!$$

is a Charlier polynomial as defined in [8]. For all c and C, the function $R_{c,C}(\tau)$ is a concave function of τ [7]. It is defined only for integer values of C, but, for all c and τ, is a strictly convex function of C in the sense that, for all U,

$$B_{c,C}(\tau, U) < S_{c,C}(\tau, U).
\tag{2}$$

In [9] a computationally stable method for efficiently calculating the functions $\widetilde{R}_{c,C}(s)$ was developed. These functions can be inverted using the Euler method [10]. Alternatively, a linear approximation of $R_{c,C}(\tau)$ itself, also developed in [9], can be used.

Similar bandwidth valuations can be derived for other traffic and link models. For example, with TCP traffic, the value of bandwidth may be expressed in terms of the effect on revenue and on user response times rather than lost connections, and models such as [11] can be used to compute the values B and S.

3   A Distributed Bandwidth Reallocation Scheme

3.1   The Logical Network

Consider a physical network consisting of nodes $n \in \mathcal{N}$ and links $\ell \in \mathcal{L}$. Assume that each link $\ell \in \mathcal{L}$ has an associated physical bandwidth B_ℓ. The network supports routes $r \in \mathcal{R}$ which form a logical overlay network. Let o_r and d_r denote the end nodes of route r.

At the level of the physical network, route r is provisioned by reserving bandwidth on a path $\mathcal{L}_r = \{\ell_1, \ell_2, \ldots, \ell_{k_r}\}$ of physical links that connects o_r to d_r. If there is a physical link connecting o_r and d_r, we use this link to provision the route, in which case $\mathcal{L}_r$ contains only one link ($k_r = 1$). Such a route is called a *direct* route. If o_r and d_r are not physically connected, then $k_r > 1$, the first node of ℓ_1 is o_r and the second node of ℓ_{k_r} is d_r. We call such a route a *transit* route and denote the set of routes, both direct and transit, that pass through link ℓ by

$$\mathcal{A}_\ell = \{r : \ell \in \mathcal{L}_r\}.$$

For a transit route r, let $\mathcal{D}_r$ be the set of direct routes that traverse the single links $\ell \in \mathcal{L}_r$. The routes in $\mathcal{D}_r$ are called the *constituent direct routes* corresponding to the transit route r.

Each route r is allocated its own bandwidth C_r such that the constraints imposed by the underlying physical network are observed. Thus, for all $\ell \in \mathcal{L}$, we must have

$$\sum_{r \in \mathcal{A}_\ell} C_r = B_\ell. \tag{3}$$

Figure 1 shows a network with three nodes and two links. Nodes n_1 and n_2 are connected by a direct route r_1 of bandwidth C_1 which is provisioned by reserving bandwidth on the link ℓ_1. Nodes n_2 and n_3 are connected by a direct route r_2 of bandwidth C_2 which is provisioned by reserving bandwidth on the link ℓ_2. Nodes n_1 and n_3 are connected by a transit route r_3 of bandwidth C_3 which is provisioned by reserving bandwidth on the links ℓ_1 and ℓ_2. Thus $\mathcal{A}_1 = \{r_1, r_3\}$ and $\mathcal{A}_2 = \{r_2, r_3\}$. The constraints (3) yield $C_1 + C_3 = B_1$ and $C_2 + C_3 = B_2$ where B_1 and B_2 denote the physical bandwidths on links ℓ_1 and ℓ_2 respectively.

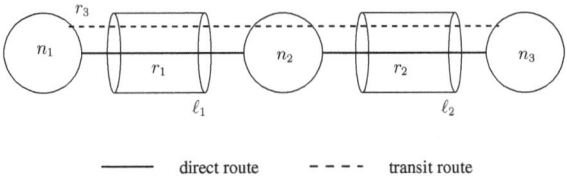

direct route   - - - -   transit route

Fig. 1. Transit and direct routes

3.2 Bandwidth Reallocation

Requests for connection on route r arrive in a Poisson stream of rate λ_r. If the current number c_r of connections on route r is less than C_r, the request is accommodated and c_r is increased by one. Otherwise the request is rejected. Successful requests hold bandwidth for a time which is exponentially distributed with mean $1/\mu_r$ (other holding time distributions are investigated in Sect. 4). The bandwidth values $B^{(r)}_{c_r,C_r}(\tau_r, U)$ and $S^{(r)}_{c_r,C_r}(\tau_r, U)$ will thus vary over time because the occupancy c_r varies over time. It is therefore likely that situations will arise where it will be advantageous to reallocate bandwidth among the routes.

As stated previously, the defining ingredient in our reallocation scheme is that bandwidth reallocation takes place between transit routes and their constituent direct routes. In the example of Fig. 1, the transit route r_3 can either acquire or release bandwidth from its constituent direct routes r_1 and r_2. If only such "local" transfers are permitted, we avoid the need to consider the potentially widespread implications of a particular reallocation.

At random instants of time, a comparison is made between the value of U extra units of bandwidth on transit route r and the sum of the costs of losing U units of bandwidth on the constituent direct routes $\ell \in \mathcal{D}_r$, and between the costs of losing U units of bandwidth on transit route r and the sum of the values of U extra units of bandwidth on the constituent direct routes $\ell \in \mathcal{D}_r$. For a transit route r, if

$$B^{(r)}_{c_r,C_r}(\tau_r, U) > \sum_{\ell \in \mathcal{D}_r} S^{(\ell)}_{c_\ell,C_\ell}(\tau_r, U), \tag{4}$$

then the transit route acquires U units of bandwidth from each of its constituent direct routes. Alternatively, if

$$S^{(r)}_{c_r,C_r}(\tau_r, U) < \sum_{\ell \in \mathcal{D}_r} B^{(\ell)}_{c_\ell,C_\ell}(\tau_r, U), \tag{5}$$

then the transit route releases U units of bandwidth to each of its constituent direct routes. If neither condition is satisfied, no reallocation occurs.

It follows from inequality (2) that the inequalities (4) and (5) cannot simultaneously be satisfied. To see this, observe that applying (2) to the direct routes in the sums on the right hand sides of the inequalities (4) and (5), we deduce that $B^{(r)}_{c_r,C_r}(\tau, U) > S^{(r)}_{c_r,C_r}(\tau, U)$, which contradicts (2) applied to the transit route r.

- At random time points, transit route r sends out a control packet that includes two pieces of "information"

$$\text{ACQUIRE} := B^{(r)}_{c_r,C_r}(\tau_r, U),$$
$$\text{RELEASE} := S^{(r)}_{c_r,C_r}(\tau_r, U).$$

- At each constituent direct route $\ell \in \mathcal{D}_r$, the information in the control packet is modified according to

$$\text{ACQUIRE} := \text{ACQUIRE} - S^{(\ell)}_{c_\ell,C_\ell}(\tau_r, U),$$
$$\text{RELEASE} := \text{RELEASE} - B^{(\ell)}_{c_\ell,C_\ell}(\tau_r, U).$$

- When the control packet reaches the destination node of the transit route r, we observe that

$$\text{ACQUIRE} := B^{(r)}_{c_r,C_r}(\tau_r, U) - \sum_{\ell \in \mathcal{D}_r} S^{(\ell)}_{c_\ell,C_\ell}(\tau_r, U),$$
$$\text{RELEASE} := S^{(r)}_{c_r,C_r}(\tau_r, U) - \sum_{\ell \in \mathcal{D}_r} B^{(\ell)}_{c_\ell,C_\ell}(\tau_r, U).$$

A check of the information is performed and
  - if ACQUIRE > 0, then $C_r := C_r + U$. A packet is sent along the reverse route instructing each constituent direct route to perform $C_\ell := C_\ell - U$, for $\ell \in \mathcal{D}_r$.
  - if RELEASE < 0, then $C_r := C_r - U$. A packet is sent along the reverse route instructing each constituent direct route to perform $C_\ell := C_\ell + U$, for $\ell \in \mathcal{D}_r$.
  - else no change occurs.

Fig. 2. An algorithm to implement the reallocation scheme

Figure 2 presents one way in which the reallocation scheme could be implemented.

3.3   Scalability, Local Information, Global Reach

The bandwidth reallocation scheme presented in Fig. 2 makes use of local information only. Thus a signal (data packet) sent along a transit route r gathers all the information needed to reallocate bandwidth on route r, namely the bandwidth valuations for route r and its constituent direct routes $\ell \in \mathcal{D}_r$.

Since the transit routes do not communicate directly with each other, the algorithm presented in Fig. 2 is distributed and scalable. The transit routes do however communicate indirectly with each other via the constituent direct routes that they have in common. For example, if a transit route r acquires/releases bandwidth then the constituent direct routes $\ell \in \mathcal{D}_r$ will release/acquire bandwidth. Other transit routes, which share these constituent direct routes, may then decide to reallocate bandwidth based upon the new valuation.

Thus, although the bandwidth reallocation scheme presented in Fig. 2 can only directly modify the values of local variables, the scheme has a global reach in the sense that the effects of local changes can spread throughout the network and influence the values of non-local variables.

3.4 Suitable Values for the Bandwidth Reallocation Parameters

The reallocation rate. In order for the reallocation scheme to be efficient, the rate of reallocation attempts should be significantly lower than the rate of connection arrivals. If this were not the case, then it would be more efficient to use connection-by-connection signalling to set up a path with sufficient bandwidth to carry each connection. We therefore assume that control packets on transit route r are transmitted in a Poisson process with rate η_r where $\eta_r = V\lambda_r$ and $V \in [0,1]$ is the *signalling ratio*. If $V = 0$ then no reallocation takes place.

The size of the signal. We explicitly take the cost of signalling into account, assigning a payload to each signal, which uses network resources. The signalling payload v is set to 64 octets/signal.

The planning horizon. The calculation of a network-wide planning horizon would either involve the use of global data (which would not scale) or require the network to enforce a common value for the planning horizon. The second of these scenarios could be implemented. However we choose to work with local (per-transit route) planning horizons τ_r that are assumed to be a multiple of the average interval $1/\eta_r$ between reallocation attempts. Thus $\tau_r = P/\eta_r$ for transit route r where $P > 0$ is the *planning ratio*, and $\tau_s = \tau_r$ for each constituent direct route $s \in \mathcal{D}_r$.

The reallocation unit. If a small unit U of bandwidth is reallocated, then very frequent reallocation might be required to keep up with the changes in bandwidth occupancy. This situation should be avoided in order to keep the overhead associated with bandwidth reallocation low. On the other hand, if a large unit of bandwidth is reallocated, the bandwidth distribution will be coarser. This situation should also be avoided in order to keep the carried traffic high.

We constructed an example network (see Sect. 4) for which we conducted simulation experiments [12] to determine suitable network-wide values for the planning ratio P, the signalling ratio V and the reallocation unit U. Based upon these experiments, we set the planning ratio P to 1, the signalling ratio V to 0.1 and the bandwidth reallocation unit U to 4. (These values are not universal but depend on the example.)

4 The Performance of the Distributed Bandwidth Reallocation Scheme

In this section we present simulation experiments to evaluate the performance of the bandwidth reallocation scheme when applied to the network model presented in Fig. 3 which represents a fictitious 30-node 46-link network based on

Fig. 3. The network model

the geography of Europe. The placement of the nodes and the links, and the construction of the traffic matrix are discussed in [12]. Each pair of cities is connected by a least cost (least hop count) route. Each route r is modelled as an $M/M/C_r$ queue. The Erlang loss formula is used to compute the amount C_r of bandwidth needed so that the route can carry its offered load ρ_r, which is obtained from the traffic matrix, with a loss probability of 2%. The figure of 2% is chosen so that the routes are initially not excessively mis-configured, yet leaving room for the reallocation method to discover a better provisioning.

Network performance is quantified in terms of the the average network loss probability and the network lost revenue. The network loss probability is given by

$$B = \sum_{r \in \mathcal{R}} \rho_r B_r / \sum_{r \in \mathcal{R}} \rho_r$$

where B_r is the connection loss probability on route r. The network lost revenue is given by

$$R = \sum_{r \in \mathcal{R}} R_r$$

where R_r is the revenue lost on route r. Let L_r denote the number of lost connections on route r. We shall assume that the revenue earned per unit time on route r is proportional to the length of the route. Hence the revenue lost on route r is $R_r = \theta_r L_r / \mu_r$ where

$$\theta_r = |\mathcal{L}_r| \theta \tag{6}$$

Table 1. Average lost revenue per 100,000 offered connections

Traffic perturbation α	Without reallocation	With reallocation	Improvement factor
0.0	6553	2953	2.2
0.1	3182	8699	2.7
0.2	4928	14677	3.0

where θ is the revenue earned per unit time on a single hop connection and $|\mathcal{L}_r|$ is the length in hops of route r.

The results of our simulations show that for the network model under investigation, bandwidth reallocation reduces the average network loss probability by a factor of 3 from 0.02 to 0.007. Table 1 shows that the average lost revenue per 100,000 offered connections is reduced by a factor of 2 from 6,553 to 2,953. Moreover, effective bandwidth reallocation can be implemented while keeping the signalling cost incurred in reallocation low.

Note that in the experiments below, increasing the link bandwidths by some 15% will yield the same reduction in lost revenue as is obtained by bandwidth reallocation. However, the purpose of this study is to determine if one mechanism (bandwidth reallocation) can deal effectively with both failure recovery, random traffic variations and time-varying traffics.

The following sections present simulation experiments which investigate the performance of the reallocation scheme. Each simulation experiment was replicated 15 times and each replication processed 10^7 connection completions.

4.1 The Response of the Scheme to Random Traffic Variations

In this section we investigate how the bandwidth reallocation scheme deals with traffic patterns which differ substantially from the traffic pattern that was originally used to configure the route bandwidths.

Let $\lambda_r(0)$ denote the original average arrival rate of connections to route r. The ith traffic perturbation yields a new set of connection arrival rates

$$\lambda_r(i) = \lambda_r(0)(1 + b_r a_r(i))$$

where $i > 0$, $a_r(i) \sim U(0, a)$ is a random variable sampled from a uniform distribution in the range $[0, a]$ and $b_r = \pm 1$ is a Bernoulli random variable with $P(b_r = 1) = 0.5$. The effect of b_r is to ensure that the connection arrival rates to a randomly selected set of the routes are increased while the arrival rates to the remaining routes are decreased. Upon the ith perturbation, the connection arrival rate to route r becomes $\lambda_r(i)$, the signalling rate on transit route r becomes $\eta_r(i) = V\lambda_r(i)$ and the planning horizon becomes $\tau_r(i) = P/\eta_r(i)$. The traffic perturbation process is repeated after every 10^6 connection completions.

Table 1 shows the average lost revenue per 100,000 offered connections. The figure shows that bandwidth reallocation substantially reduces the lost revenue when the network traffic is subject to repeated random perturbations of $\pm 10\%$ and $\pm 20\%$. The network cannot cope with these traffic perturbations if bandwidth reallocation is not used.

4.2   The Response of the Scheme to Time-Varying Traffics

In this section we investigate the efficacy of the bandwidth reallocation scheme when the traffics are time-dependent. We will assume that the average connection arrival rates are subject to a sinusoidal variation which could model a daily traffic pattern.

Let N denote the number of connection completions simulated. Let $\Lambda = \sum_r \lambda_r$ denote the total connection arrival rate to the network. Let $T = N/\Lambda$ denote the average time required to complete N connections. A sinusoidal variation

$$\lambda_r(t) = \lambda_r(1 + a\sin(\omega(\phi_r + t))) \qquad (7)$$

is applied to the connection arrival rates λ_r to each route $r \in \mathcal{R}$ where $a \geq 0$ is an amplitude factor, $\omega = 2\pi/T$, t is the current simulation time and $\phi_r = b_r\, T/2$ where $b_r = \{0,1\}$ is a Bernoulli random variable with $P(b_r = 1) = 0.5$. The effect of ϕ_r is to ensure that the arrivals to a randomly selected set of the routes are π radians out of phase with respect to the arrivals to the remaining routes. At any time in the simulation, the connection arrival rates $\lambda_r(t)$ to a randomly chosen set of the routes are such that $\lambda_r(t) \geq \lambda_r(0)$ while the arrival rates to the remaining routes are such that $\lambda_r(t) \leq \lambda_r(0)$. The signalling rate on transit route r becomes $\eta_r(t) = V\lambda_r(t)$.

Figure 4 shows the point estimates of the average lost revenue per 100,000 offered connections. The figure shows that bandwidth reallocation substantially reduces the lost revenue when the network traffic is subject to a sinusoidal variation of $\pm 10\%$ and $\pm 20\%$. The network cannot cope with such time-varying traffic if bandwidth reallocation is not used.

4.3   The Effect of the Connection Holding Time Distribution

This section presents results from simulation experiments which test the robustness of the bandwidth reallocation scheme with respect to the assumptions that were used to derive the bandwidth value function in [7].

Figure 5 presents the total lost revenue when the holding times are not exponentially distributed (recall that the derivation of the value function assumes that the holding times are exponentially distributed). Specifically, we compare performance when the simulated holding times have log normal, Erlang K and Pareto distributions. In each case a total of 10^8 connection completions were simulated. The parameterization of the holding time distributions is discussed in [12].

Figure 5 shows that the bandwidth reallocation scheme substantially reduces the lost revenue in spite of the mismatch between the simulated holding time distribution and the holding time distribution assumed by the bandwidth value function. The reallocation scheme remains effective as the squared coefficient of variation is varied over five or three or two orders of magnitude for the lognormal or Erlang-k or Pareto holding time distributions respectively.

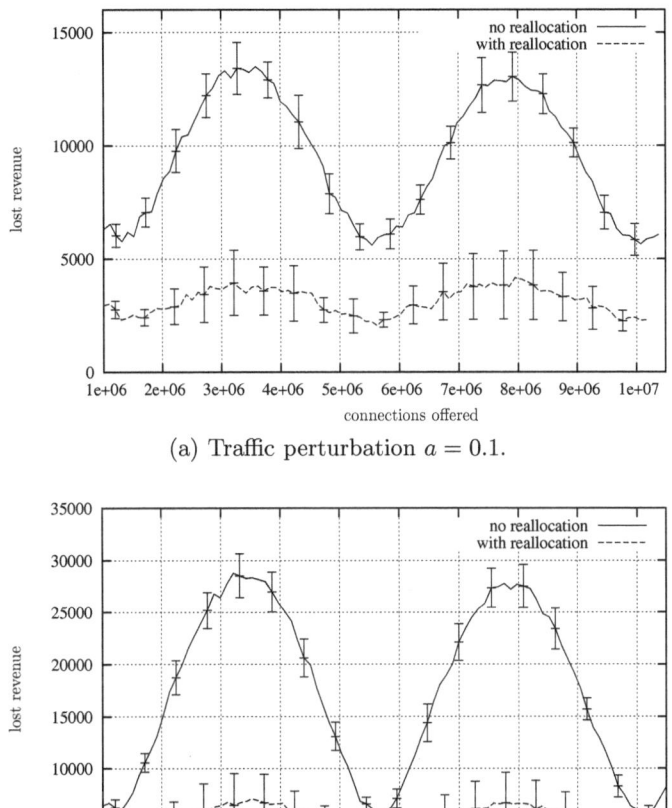

(a) Traffic perturbation $a = 0.1$.

(b) Traffic perturbation $a = 0.2$.

Fig. 4. The lost revenue when the connection arrival rates are varied sinusoidally

4.4   The Effect of Errors in the Traffic Estimates

The manager of each route r needs to know the connection arrival rate λ_r, the signalling rate η_r, the connection holding time $1/\mu_r$ and the number c_r of connections in service in order to calculate the bandwidth valuations. In our simulation experiments, we assumed that the values of these parameters were known and we used them in our calculations. In a real implementation they would have to be estimated on-line and errors may arise in their estimation as in for example [13].

In this section we investigate the effect of errors in the estimated value of the connection arrival rate on the efficacy of the reallocation scheme. Let $\lambda'_r = \lambda_r(1 + \delta_r)$ denote an estimate of λ_r where δ_r is a random variable sampled

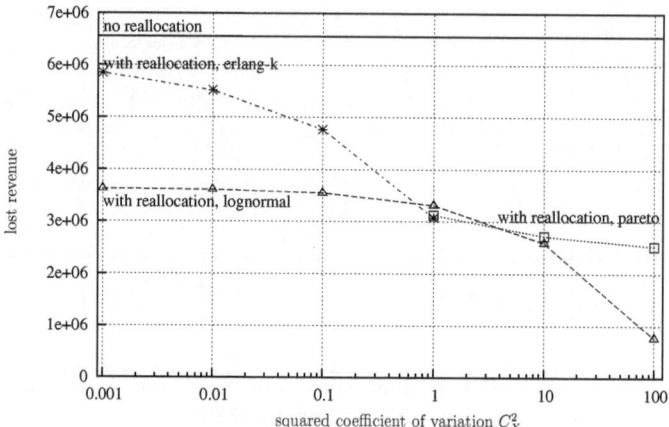

Fig. 5. The robustness of the reallocation scheme with respect to the connection holding time distribution

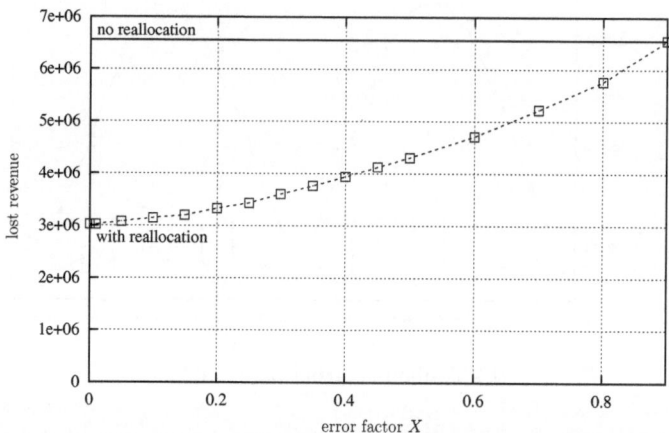

Fig. 6. The reallocation scheme is relatively insensitive to errors in the estimation of the connection arrival rates

from a uniform distribution in the range $[-X, X]$ where $X < 1$. A value of λ'_r is computed by sampling a new value of δ_r whenever the reallocation scheme computes the value of the bandwidth on route r.

Figure 6 presents the total lost revenue for several values of the arrival rate error parameter X. A total of 10^8 connection completions were simulated. Confidence intervals are omitted since they are so small that they are indistinguishable from the plot itself. The figure shows that the reallocation scheme is relatively insensitive to reasonable errors in the estimated values of the connection arrival rates. This result is in agreement with [14] where it was shown that it is possible

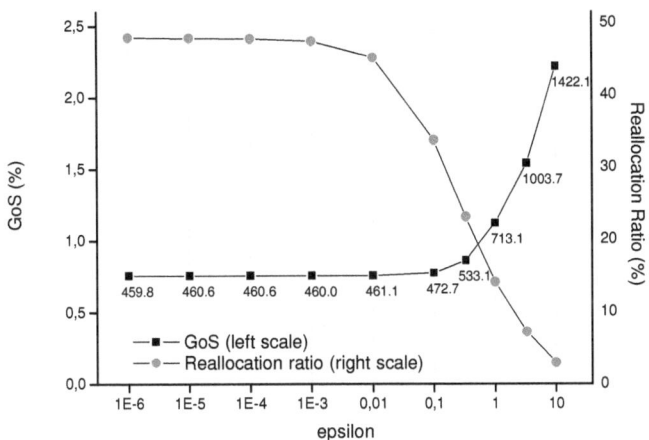

Fig. 7. The reallocation scheme is relatively insensitive to the floor price

to obtain a robust routing that guarantees a nearly optimal utilization with a fairly limited knowledge of the applicable traffic demands.

4.5 Floor Price Sensitivity

The floor price ϵ denotes the minimum reduction in the expected lost revenue that is necessary for a bandwidth reallocation to take place. Although an optimum value for the floor price can be determined experimentally, there would be no need to find a near-optimal value for the floor price if the reallocation scheme were relatively insensitive to the value of the floor price.

Figure 7 presents the GoS, the revenue loss (indicated by the numerical annotations on the GoS plot) and the reallocation ratio for several values of the floor price ϵ. Confidence intervals are omitted since they are sufficiently small to be indistinguishable from the plot itself. The figure shows that the reallocation scheme is relatively insensitive to the value of the floor price. Note that a larger value for the floor price will reduce the number of successful reallocations. This will lead to a decrease in the reallocation success ratio which is defined as the number of successful reallocations versus the total number of reallocations attempted. This ratio quantifies the wasted effort spent in gathering pricing information for capacity reallocation attempts which, if they were implemented, would result in revenue gains which are less than the floor price.

4.6 The Effect of Signalling Delays

In this section we investigate the effect of signalling delays on the efficacy of the bandwidth reallocation scheme. Excessive signalling delays will cause the reallocation scheme to work with out-of-date information which might give rise to inappropriate bandwidth reallocations and mis-configured routes.

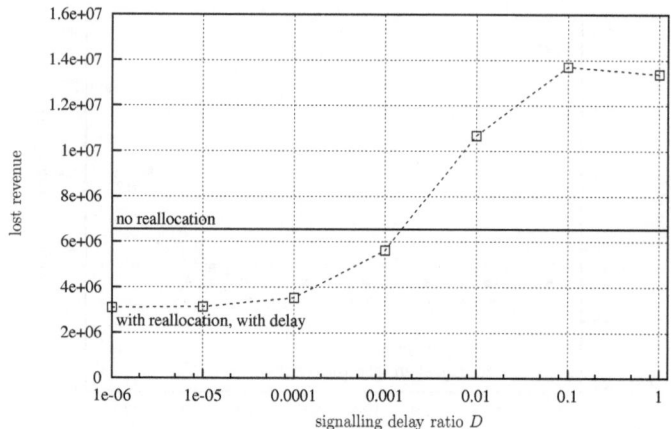

Fig. 8. The lost revenue versus the link signalling delay ratio D

The signalling information is transported in control packets which we assume are transmitted without loss. The loss of control packets would require timeouts to perform retransmissions of lost packets or cancellation of reserved but not yet committed bandwidth resources.

The signalling delay on a transit route is the sum of the delays experienced on each link of the transit route. The link delay, which is the sum of the link propagation, link transmission and nodal processing delays (including the computation of bandwidth price values), is modelled as a random variable sampled from an exponential distribution with parameter κ. The average signalling delay $1/\kappa$ per link is expressed as a fraction of the average connection holding time $1/\mu$ so that $1/\kappa = D/\mu$. Thus $D = 10^{-5}$ for a connection with a 15 minute holding time implies an average signalling delay of 9 msecs per link.

Figure 8 plots the total lost revenue versus the link signalling delay ratio D. A total of 10^8 connection completions were simulated. The scale on the D axis is logarithmic. The figure shows that in the region of interest where $D < 10^{-4}$, the link signalling delays have minimal impact on the reallocation scheme. However, in the region $D > 10^{-2}$ where the signalling delays are excessively long, many bandwidth reallocation decisions are taken based on out-of-date bandwidth value information. In this region the route capacities become mis-configured to such an extent that the lost revenue exceeds that observed when no reallocations are attempted.

4.7   Bandwidth Reallocation among Service Classes

In this section we investigate another aspect of the robustness of the bandwidth reallocation scheme where the scheme is used to protect the GoS of one traffic *class* against the bandwidth demands of another traffic class.

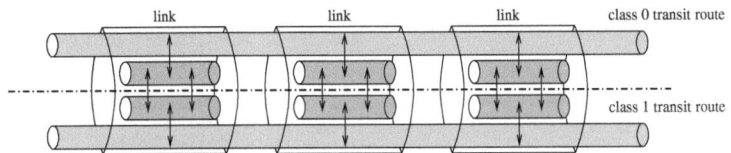

Fig. 9. Inter-class trading: transit routes trade with each other via their respective constituent direct routes

The calls offered to the network are partitioned into two classes referred to as class 0 and class 1. Two routes connect each OD-pair, one route for class 0 traffic and the other route for class 1 traffic. We consider two forms of bandwidth reallocation, namely *intra-class* and *inter-class* reallocation. In intra-class reallocation, bandwidth reallocation takes place between a transit route and its constituent direct routes, and no reallocation takes place between the class 0 and class 1 constituent direct routes. In this case, the class 0 and class 1 virtual networks are managed independently.

In inter-class reallocation, bandwidth reallocation takes place between a transit route and its constituent direct routes, and in addition reallocation takes place between the constituent direct routes of the different classes as illustrated in Fig. 9. Bandwidth reallocation between the constituent direct routes of the different classes is referred to as *cross-trading*.

The arrival rates $\lambda_r(t)$ of class 1 calls are subjected to a time-dependent sinusoidal variation. The arrival rates λ_r of class 0 calls are left unchanged. Perturbing the arrival rates of class 1 calls only allows us to compare the efficiency of intra-class bandwidth reallocation in the presence or absence of traffic variations. In addition, it will be shown that inter-class trading improves the network's overall GoS, although the GoS of the unperturbed class 0 may suffer when its allocated bandwidth is acquired by the perturbed class 1 in order to better accommodate its varying traffic.

The network model presented in Fig. 3 is used to investigate the effect of inter- and intra-class reallocation. For computational efficiency, the approximate pricing scheme in [9] is used instead of the pricing scheme in [7]. Figure 10 presents the GoS with intra-class trading. Bandwidth reallocation improves the GoS of the unperturbed class 0 from 2.09% to 0.75%. The arrival rates of class 1 connections are subject to a sinusoidal perturbation with random phase: see (7) where $\phi_r = b_r T/2$ and b_r is a uniform random variable in the range $(0, 1)$. The GoS of class 1 is reduced from 5.67% to 1.58%. This demonstrates that the reallocation scheme can provide effective bandwidth assignment despite the traffic perturbation.

Figure 10 also presents the GoS when inter-class (cross-trading) bandwidth reallocation is enabled. At the expense of a small deterioration in the GoS of the non-perturbed class 0 from 0.75% to 0.94%, the GoS of the perturbed class 1 improves from 1.58% to 1.04%.

Although inter-class trading has a beneficial impact on network performance, class 0 is now indirectly exposed to the traffic fluctuations of class 1 since class 1

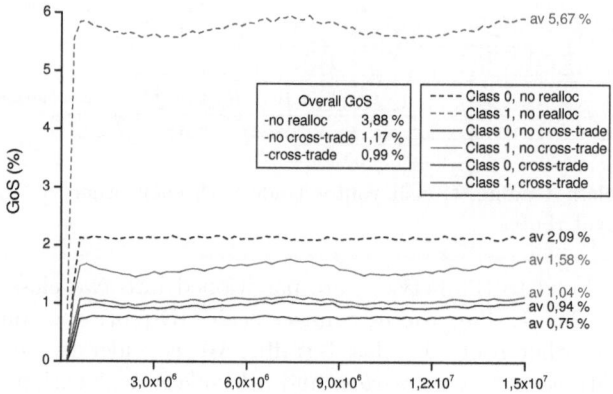

Fig. 10. The results of inter-class trading: the GoS for the unperturbed class 0 and the sinusoidally perturbed (random phases) class 1

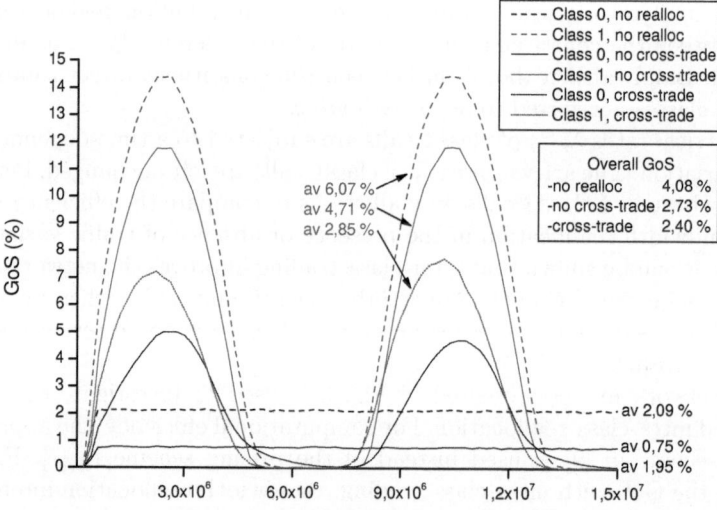

Fig. 11. The results of inter-class trading: the GoS for the unperturbed class 0 and the sinusoidally perturbed (constant phases) class 1

can acquire bandwidth from the constituent direct routes of its non-perturbed counterpart. Thus only the perturbed traffic of class 1 strictly benefits.

The impact of inter-class trading is further examined in Fig. 11. As before, the network carries two traffic classes. Class 0 is not perturbed and class 1 is subject to a synchronised sinusoidal perturbation. Figure 11 presents the GoS of both call classes when no bandwidth reallocation takes place. The figure shows that that the GoS of both classes is substantially improved when intra-class

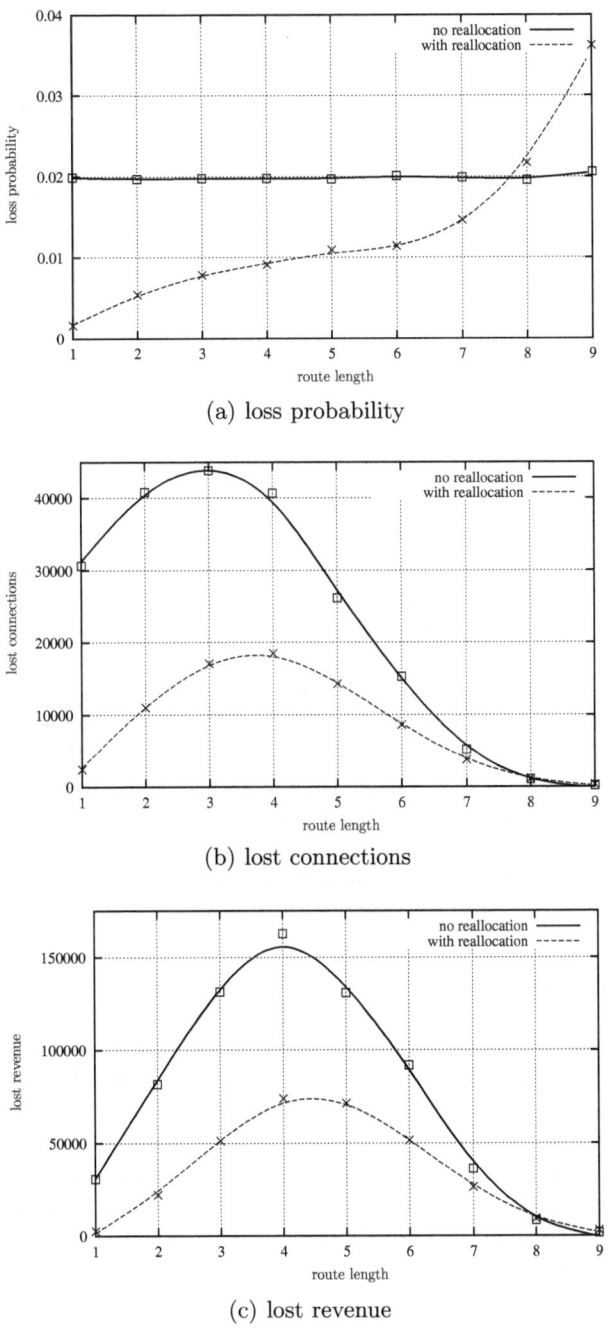

(a) loss probability

(b) lost connections

(c) lost revenue

Fig. 12. The loss probability, the number of lost connections and the lost revenue as a function of the route length

bandwidth reallocation takes place. The figure shows that the GoS of both classes improves further when inter-class bandwidth reallocation takes place. However, the inter-class trading allows the indirect reallocation of bandwidth between the two classes and the sinusoidal perturbation of class 1 is communicated to the unperturbed class 0.

4.8 Network Performance as a Function of Route Length

In a network with complete sharing of bandwidth, a connection is carried only if sufficient bandwidth is available on every link of the route. This can cause discrimination against long routes because the conditions for a connection to be admitted on a long route are more stringent than they are on shorter routes. This problem is overcome when we allocate bandwidth end-to-end, because the link-by-link test for admission is avoided. However, there is the potential for the problem to be reintroduced by a bandwidth reallocation scheme such as the one that we have proposed here. This is because it might be difficult for a long route to obtain the bandwidth that it needs from a large number of constituent direct routes. In this section we examine whether this potential problem occurs in our test network with the parameter values as defined above.

Figure 12 shows the loss probability, the total number of lost connections and the total lost revenue for routes of length m where $1 \leq m \leq 9$. The figure shows that reallocation substantially reduces the loss probabilities, the lost connections and the lost revenue for all but the longest routes. Note that the discrimination against long routes is not inherent to the bandwidth reallocation scheme: it is rather a consequence of the value assigned to the weight θ_r in (6). Although the loss probabilities on routes of lengths 8 and 9 are high, the number of lost connections and the lost revenue on these routes is negligible. The concave shape of the lost connection and the lost revenue curves as a function of the route length m is due to the route length distribution and the fact that the revenue earned on a route r is proportional to the length of the route.

5 Conclusions

This paper presents a distributed scheme for bandwidth reallocation in path-oriented transport networks. The defining feature of the scheme is that bandwidth reallocation takes place between transit routes and their constituent direct routes. In this way the scheme is autonomous, acts without centralized control from a system coordinator and behaves entirely according to local rules.

We have presented a series of simulation experiments implementing a version of such a bandwidth reallocation scheme. A model of a fictitious 30-node 46-link network is investigated. The simulation studies reveal that bandwidth reallocation can substantially reduce the revenue lost by the network due to short time-scale random fluctuations and to long time-scale changes in traffic arrival rates.

Although we have not investigated any mechanisms to tune the scheme, we have presented some insight into how a number of the parameters of the scheme should be chosen. Our results are consistently positive and indicate that schemes within the general framework that we have presented here deserve further investigation.

References

1. Krzesinski, A., Müller, K.: A distributed scheme for bandwidth re-allocation between working and recovery paths. In: South African Institute for Computer Scientists and Information Technologists SAICSIT 2006, South Africa, October 2006, pp. 135–144 (2006)
2. Liu, Y., Tipper, D., Siripongwutikorn, P.: Approximating optimal spare capacity allocation by successive survivable routing. IEEE/ACM Transactions on Networking 13(1), 198–211 (2005)
3. Lanning, S., Massey, W., Rider, B., Wang, Q.: Optimal pricing in queueing systems with quality of service constraints. In: 16th International Teletraffic Congress, pp. 747–756 (1999)
4. Fulp, E., Reeves, D.: QoS rewards and risks: A multi-market approach to resource allocation. In: Pujolle, G., Perros, H.G., Fdida, S., Körner, U., Stavrakakis, I. (eds.) NETWORKING 2000. LNCS, vol. 1815, pp. 945–956. Springer, Heidelberg (2000)
5. Kelly, F., Mualloo, A., Tan, D.: Rate control for communication networks: Shadow prices, proportional fairness and stability. Journal of the Operational Research Society 49, 237–252 (1998)
6. Low, S., Lapsley, D.: Optimization flow control, I: Basic algorithm and convergence. IEEE/ACM Trans. on Networking 7(6), 861–874 (1999)
7. Chiera, B., Taylor, P.: What is a unit of capacity worth? Probability in the Engineering and Informational Sciences 16(4), 513–522 (2002)
8. Chihara, T.: An Introduction to Orthogonal Polynomials. Gordon and Breach, Science Publishers Inc., Aachen, Germany (1978)
9. Chiera, B., Krzesinski, A., Taylor, P.: Some properties of the capacity value function. SIAM Journal on Applied Mathematics 65(4), 1407–1419 (2005)
10. Abate, J., Whitt, W.: Numerical inversion of laplace transforms of probability distributions. ORSA Journal on Computing 7(1), 36–43 (1995)
11. Arvidsson, Å., Krzesinski, A.: A model for TCP traffic. In: 15th International Teletraffic Congress Specialist Seminar, pp. 68–77 (2002)
12. Arvidsson, Å., Chiera, B., Krzesinski, A., Taylor, P.: A simulation of a bandwidth market. In: Technical Report, Department of Computer Science. University of Stellenbosch, 7600 Stellenbosch, South Africa (2006)
13. Arvidsson, Å.: High level B-ISDN/ATM traffic management in real time. In: Kouvatsos, D. (ed.) Performance Modelling and Evaluation of ATM Networks, vol. 1, pp. 177–207. Chapman & Hall, London (1994)
14. Applegate, D., Cohen, E.: Making intra-domain routing robust to changing and uncertain traffic demands: understanding fundamental tradeoffs. In: Conference on Applications, Technologies, Architectures, and Protocols for Computer Communications, Karslruhe, Germany, December 2003, pp. 313–324 (2003)

A Fair and Dynamic Load-Balancing Mechanism

Federico Larroca and Jean-Louis Rougier

Télécom ParisTech, Paris, France
46 rue Barrault F-75634 Paris Cedex 13
firstname.lastname@telecom-paristech.fr

Abstract. The current data network scenario makes Traffic Engineering (TE) a very challenging task. The ever growing access rates and new applications running on end-hosts result in more variable and unpredictable traffic patterns. By providing origin-destination (OD) pairs with several possible paths, load-balancing has proven itself an excellent tool to face this uncertainty. Most previous proposals defined the load-balancing problem as minimizing a certain network cost function of the link's usage, assuming users would obtain a good performance as a consequence. Since the network operator is interested in the communication between the OD nodes, we propose instead to state the load-balancing problem in their terms. We define a certain utility function of the OD's perceived performance and maximize the sum over all OD pairs. The solution to the resulting optimization problem can be obtained by a distributed algorithm, whose design we outline. By means of extensive simulations with real networks and traffic matrices, we show that our approach results in more available bandwidth for OD pairs and a similar or decreased maximum link utilization than previously proposed load-balancing schemes. Packet-level simulations verify the algorithm's good performance in the presence of delayed and inexact measurements.

1 Introduction

Network convergence is a reality. Many new services such as P2P or HD-TV are offered on the same network, increasing the unpredictability of traffic patterns. To make matters worse, access rates have increased at such pace that the old assumption that core link capacities are several orders of magnitude bigger than access rates is no longer true. Thus, simply upgrading link capacities may not be an economically viable solution any longer. This means that network operators are now, more than ever, in need of Traffic Engineering (TE) mechanism which are **efficient** (make good use of resources), but also **automated** (as much self-configured as possible), more **robust** with respect to network variations (changes in traffic matrix, or characteristics of transported flows) and more **tolerant** (in case of node/link failures).

Dynamic load-balancing [1, 2, 3] is a TE mechanism that meets these requirements. If an origin-destination (OD) pair is connected by several paths, the problem is simply how to distribute its traffic among these paths in order to achieve a certain objective. In these dynamic schemes, paths are configured a

R. Valadas and P. Salvador (Eds.): FITraMEn 2008, LNCS 5464, pp. 36–52, 2009.
© Springer-Verlag Berlin Heidelberg 2009

priori and the portion of traffic routed through each of them (traffic distribution) depends on the current traffic matrix (TM) and network's condition. As long as the traffic distribution is updated frequently enough, this kind of mechanism is **robust** and their dependence on the network's condition makes them naturally **tolerant** too. Finally, if the algorithm is also distributed (in the sense that each router makes its choices independent of the others) the resulting scheme will also be **automated**.

In intra-domain TE, the network operator is interested in the communication between the OD nodes, i.e. the performance they get from their paths. The OD pairs may actually be regarded as the users of the network, sharing its resources between them. It is natural then to state the load-balancing problem (or TE in general) in their terms. An analogy can be made with the congestion control problem [4], where the users are the end-hosts and share link capacities. The user's performance (or "revenue") is the obtained rate and the objective is to maximize the sum over all users of a utility function of it. In our case the problem is different since the rate is given and we control only the portion of traffic sent through each path. In this paper we propose to measure the user's performance by the mean available bandwidth (ABW) the OD pair obtains in its paths, and then maximize the sum over all pairs of a utility function of this measure. We will present a distributed algorithm in which the independent adjustments made by each OD pair lead to the global optimum. Our comparison with previously proposed load-balancing schemes, using several real networks and TMs, shows that the resulting traffic distribution improves OD pairs' perceived performance and decreases maximum link utilization.

Almost all prior proposals in load-balancing (and in TE in general) define a certain link-cost function of the link's capacity and load, and minimize the total network's cost defined as the sum over all links of this function. The resulting traffic distribution will be relatively balanced, in the sense that no single link will be extremely loaded. However, it is not the situation of isolated links, but the condition on the complete path(s) connecting OD nodes that counts. Solving the problem in terms of the links is only an indirect way of proceeding which does not allow us, for instance, to prioritize a certain OD pair or to enforce fairness among the OD pairs. For example, consider the network in Fig. 1. In it, all link capacities are equal and all sources generate the same amount of traffic. However, only OD pair 1 has more than one path to choose from. It is relatively simple to verify that if the link-cost function is the same in all links, the optimum is obtained when traffic from OD pair 1 is equally distributed among paths. However, since the upper path "disturbs" two OD pairs while the lower one disturbs only one, depending on our fairness definition it could make more sense to route more traffic from OD pair 1 through the lower path.

The rest of the paper is organized as follows. The following section discusses related work. Section 3 defines the network model and associated notation, followed by the presentation of the utility objective function. In Sec. 4 we address the resolution of the problem. We present some flow-level simulations in Sec. 5, where we show the performance of the distributed algorithm and the advantages of our

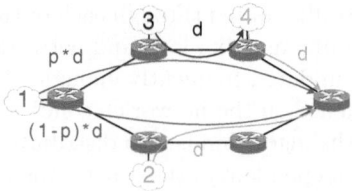

Fig. 1. An example in which if the total cost of the network is minimized the resulting optimum is unfair ($p_{opt} = 0.5$)

scheme over other TE techniques. In Sec. 6 we discuss some implementation issues, and present some packet-level simulations. We conclude the paper in Sec. 7.

2   Related Work

Load-Balancing can be seen as a particular case of Multi-Path routing. Many papers fall under this category, but we will highlight only those closely related to our work, or which inspired us for our proposal. We will further classify research in this topic into two sub-categories depending on the time-scale under consideration.

2.1   Long Time-Scale

Here we consider the minute or hour time-scale of routing. A TM is assumed to exist and is used as an input to the problem. At this time-scale congestion control is not considered, i.e. the TM is assumed independent of the current condition of the network. However, it is clear that the TM is not static and changes over time [5]. Furthermore, some network operators offer their customers VPN services in which they only have to specify the total maximum rate each node can receive or send [6]. Load-Balancing has proved itself a very effective tool to cope with this TM uncertainty, and research in this area differs mainly in choosing which uncertainty set contains the real TM at all times. Robust routing considers a well-defined set, while dynamic load-balancing only the current TM.

   The objective in robust routing is to find a unique static routing configuration that fulfills a certain criteria, generally the one that minimizes the maximum link utilization over all TMs of the corresponding uncertainty set. The set can be for instance the TMs seen the previous day, or the same day the previous week [7]. A very used option is the polytope proposed in [8] which allows for easier and faster optimization. In any case, since a single traffic distribution that works for all TMs is used, resources will be wasted for any specific TM. Shrinking the uncertainty set results in improved performance, and there are some papers in this direction [9, 10]. This shrinking should be carefully done though, because if the selected set is too small and the network faces an unforeseen TM, the resulting performance is unpredictable. Finally, optimizing under uncertainty is more difficult than "normal" optimization. This increased difficulty forces the use of simpler optimization criteria which can lead to a not so good performance (e.g. it is known that minimizing the biggest link utilization generally results in the use of longer paths).

In dynamic load-balancing, each origin node estimates its entries in the TM and, based on feedback from the network, adapts the portion of traffic it sends through each path. After some iterations, and if the TM does not change in the meantime, a global optimum is achieved. The two most well-known proposals in this area are MATE and TeXCP. In MATE [1], a convex link cost function is defined, which depends on the link's capacity and load. The objective is to minimize the total network cost, for which a simple gradient descent method is proposed. TeXCP [2] proposes a somewhat simpler objective: minimize the biggest utilization (ρ_l/c_l) each OD pair sees in its paths. A rough description of the algorithm is that origin nodes iteratively increase the portion of traffic sent through the path with the smallest utilization. Another load-balancing scheme which has the same objective but a relatively different mechanism is REPLEX [3].

2.2   Short Time-Scale

This short time-scale refers to the congestion control time-scale. Possible adaptations of TCP to the multi-path case (MP-TCP) have been extensively studied, where the utility each user perceives is now a function of the total rate he obtains from all his paths. Several propositions exist in this direction. For instance, in [11,12,13] the user is responsible of calculating his total rate and how much he should send through each path. In [14], the user only calculates the total sending rate and the routers distribute traffic among paths.

A different but related problem is a user downloading the same file from different sites or hosts (as in Bittorrent). Currently, greedy policies are used where users change a path only if they obtain a better performance on the new one. In [15] the authors show that if current TCP flavors are used in such schemes, the resulting allocation can be both unfair and inefficient, and that a mechanism similar to MP-TCP should be used instead.

In [16] the objective is to adapt the sending rates to maximize the total users' utility minus a network cost. The idea is that users should also take into account the utilization of the links and leave a margin for future arrivals. We believe that this is not the best criteria. Congestion control should enable users to consume *all* their fair-share of the path. At this time-scale, saving a little bandwidth for future arrivals is, in our opinion, a waste of resources.

Although MP-TCP constitutes a very interesting long-term objective, no actual implementations of it exists. Allowing end-hosts to choose their paths, or even making them aware that several possibilities exist, presents several technical difficulties in current Internet architectures.

3   Source-Level Utility Maximization

3.1   Network Model

We represent the network as L unidirectional links, indexed by l, whose capacities are given by the column vector $c = [c_1 \dots c_L]^T$. We will reference OD pairs by index $s = 1 \ldots S$. By abuse of notation we will also reference by s its *source node*, defined as the router through which its traffic ingress the network. This node will be in charge of distributing this traffic among paths, and in the sequel we will use the terms source and OD pair without differentiation. Each source s has n_s possible paths towards its destination, indexed by i. R_s is a $L \times n_s$ matrix whose li entry is 1 only if source s uses link l in its i-th path, and 0 otherwise. The incidence matrix is then $R = [R_1 \ldots R_S]$.

All traffic in the network is assumed to be elastic (i.e. controlled by TCP). We suppose that flows arrive to source s as a Poisson process of intensity λ_s. Each of these flows consist of a random arbitrarily distributed workload (with mean ω_s) they want to transfer, generating a demand $d_s = \lambda_s \omega_s$. Each flow is routed through path P_{si} with probability p_{si}, and it uses it throughout its lifetime. It is worth noting that we consider a dynamic context, in which flows appear and have a finite lifetime. It is also important to highlight that we are enforcing flow-level load-balancing. Packet-level load-balancing (where packets from the same flow can take different paths) may have a negative impact on TCP performance due to packet reordering on the receiver's side.

The demand on path si is then $p_{si}d_s = d_{si}$. The traffic distribution is defined simply as $d = [d_{11} \ldots d_{1n_1} \ldots d_{S1} \ldots d_{Sn_S}]^T$, and the total load on link l (ρ_l) can be easily calculated as the l-th entry of $R \times d$. Under these assumptions, if ρ_l is strictly smaller than c_l for all l, the number of flows in the network will not go to infinity [17], meaning that the network supports the given traffic distribution.

3.2   The Utility Function

Since we consider the OD pairs as the users of the network, a single performance indicator per pair should be used. Even if we considered end-hosts as the users, a single indicator per OD pair is also adequate since traffic belonging to a given OD pair is composed of many flows generated by several end-hosts.

Our proposal defines first a revenue function $u_s(d)$ which indicates the performance perceived by source s when the traffic distribution is d. The question is how this revenue should be distributed among sources. We could for instance maximize the average or the smallest of them. Drawing on the work on congestion control [4], we define a concave non-decreasing utility function $U_s(u_s)$ that represents the satisfaction source s has with its revenue $u_s(d)$, and maximize the sum over all sources. The problem in this most general version reads like this:

$$\underset{d}{\text{maximize}} \sum_{s=1}^{S} d_s U_s(u_s(d)) \qquad (1)$$

$$\text{subject to } Rd < c,\ d \geq 0 \text{ and } \sum_{i=1}^{n_s} d_{si} = d_s$$

We multiply each utility by d_s to give more weight to those nodes generating more traffic. The constraints assure that the number of living flows is finite, that there are no negative demands and that all traffic is routed.

A typical example of $U(x)$ is the utility function that leads to the so-called α-fairness [18]:

$$U(x) = \begin{cases} (1-\alpha)^{-1}x^{1-\alpha}, & \alpha \neq 1 \\ \log(x), & \alpha = 1 \end{cases} \tag{2}$$

Throughout our simulations we will use $\alpha = 1$, which results in proportional fairness [19].

Probably the most delicate part of the problem is defining $u_s(d)$. A relatively simple path performance measure is its available bandwidth (ABW). The ABW of path si is defined as $ABW_{si} = \min_{l \in si}\{c_l - \rho_l\}$. The meaning of this indicator is twofold. On the one hand, it is a rough estimator of the throughput TCP flows will obtain from the path [20, 21]. On the other hand, a path with a big ABW is a "healthy" path, in the sense that it can accommodate future unexpected increases in traffic. Our definition for u_s will be the average ABW seen by source s in all its paths, which presents a good balance between current good conditions and prudence. Substituting u_s in (1) results in:

$$\underset{d}{\text{maximize}} \sum_{s=1}^{S} d_s U_s \left(\sum_{i=1}^{n_s} p_{si} \min_{l \in si}\{c_l - \rho_l\} \right) \tag{3}$$

This version of problem (1) is very important for the elastic traffic case. Although TCP takes care of path's resource sharing, routing constitutes a degree of freedom in the obtained rate that may be taken into account. Since the mean obtained rate depends on the amount of traffic the flow is sharing its path with, this obtained rate may be indirectly controlled through routing. Let us assume that this relation is simply that TCP flows traversing path si achieve a mean rate equal to ABW_{si}. Then problem 3 is very similar to the multi-path TCP one (see Eq. 4 in [15]) where each OD pair is seen as serving d_s MP-TCP flows. Notable differences are that the decision variable is the *portion* of traffic sent through each path (and not the *amount* of traffic), and that the mean of the flow rate is used. However, by using standard TCP and changing the ingress nodes only (not all end-hosts), users can now be regarded *as if* they were using MP-TCP, with all the advantages that this means (better performance and more supported demands).

4   Solving the Problem: Distributed Algorithm

As the network's size increases, a centralized algorithm that solves (3) for the current TM does not scale well. In this context, a distributed algorithm is not only desirable but, if the network's size is considerable, necessary. In this section we present a distributed solution for our problem.

Let us rewrite (3) introducing the auxiliary variable t_{si}:

$$\underset{d,t}{\text{maximize}} \sum_{s=1}^{S} d_s U_s \left(\sum_{i=1}^{n_s} p_{si} t_{si} \right) \qquad (4)$$

$$\text{subject to} \quad t_{si} \le c_l - \rho_l \quad \forall s,i \ \forall l : l \in si$$

$$t_{si} > 0 \ , \ d \ge 0 \ \text{and} \ \sum_{i=1}^{n_s} d_{si} = d_s$$

Although all constraints are affine, the objective function is not concave, meaning that methods solving the dual problem will only find a lower bound for the optimum. How tight is this lower bound (i.e. how small is the duality gap) is closely related to the lack of concavity of the function [22]. Our estimations indicate that in this case the lack of concavity is relatively small and decreases with the number of paths. In view of these results, we applied the well-known Arrow-Hurwicz method [23] and confirmed that the resulting traffic distribution is a very tight approximation to the optimum.

This method is iterative and at each step updates the value of the dual (primal) variables moving them in the direction of (opposite to) the gradient of the Lagrangian function. In this case, the Lagrangian function is:

$$L(p,t,\theta) = - \sum_{s=1}^{S} d_s U_s \left(\sum_{i=1}^{n_s} p_{si} t_{si} \right) + \sum_{s=1}^{S} \sum_{i=1}^{n_s} \sum_{l:l \in si} \theta_{sil}(t_{si} - c_l + \rho_l) \qquad (5)$$

Paths with zero ABW will not be used by the algorithm and the conditions on p_{si} will be necessarily true (normalization should be carefully done, though). Since the constraints are enforced by the algorithm (presented below), we omitted the Lagrange multipliers associated with the positiveness of p_{si} and t_{si}, and the normalization condition on p_{si}. However, θ_{sil} plays a very important role since it represents the cost of link l generated by source s in its path i, resulting in a total cost of $\widehat{\theta}_l = \sum_{s=1}^{S} \sum_{i:l \in si} \theta_{sil}$.

The derivatives of (5) with respect to $p_{s_0 i_0}$ and $\theta_{s_0 i_0 l_0}$ are:

$$\frac{\partial L}{\partial p_{s_0 i_0}} = -d_{s_0} U'_{s_0} \left(\sum_{i=1}^{n_{s_0}} p_{s_0 i} t_{s_0 i} \right) t_{s_0 i_0} + d_{s_0} \sum_{l \in s_0 i_0} \widehat{\theta}_l$$

$$\frac{\partial L}{\partial \theta_{s_0 i_0 l_0}} = t_{s_0 i_0} - c_{l_0} + \rho_{l_0}$$

The auxiliary variable t_{si} does not have much physical meaning, except that for any given p its optimal value is ABW_{si}. The derivative on θ_{sil} does not tell us much then, except that it should decrease when l is not the bottleneck of si (meaning that in such case its value should tend to zero). This forces us to estimate the value of θ_{sil}. Before discussing possible estimations, we will present the distributed algorithm:

Link algorithm. At times $t = 1, 2, \ldots$ link l:

1. Receives path demands $d_{si}(t)$ from all sources using it, and estimates its total load $\rho_l(t)$.
2. Computes its cost for each path $\theta_{sil}(t)$ and its total cost $\widehat{\theta}_l(t)$.
3. Communicates this last value and its ABW to all traversing sources.

Source algorithm. At times $t = 1, 2, \ldots$ source s:

1. Estimates its current demand $d_s(t)$.
2. Receives from the network the cost $\widehat{\theta}_l(t)$ of all links it uses and their ABW.
3. Computes the available bandwidth of its paths $(ABW_{si}(t))$ and its mean ABW $(u_s(t))$.
4. For each of its paths, it calculates the number:

$$\Delta_{si}(t) = d_s(t) U_s' \left(u_s(t) \right) ABW_{si}(t) - d_s(t) \sum_{l \in si} \widehat{\theta}_l(t)$$

5. It finds the path i_{max} with the biggest $\Delta_{si}(t)$ $(\Delta_s^{max}(t))$. It then updates each p_{si} in the following manner (where γ is a small constant):

$$p_{si}(t+1) = [p_{si}(t) + \gamma(\Delta_{si}(t) - \Delta_s^{max}(t))]^+$$

$$p_{si_{max}}(t+1) = 1 - \sum_{\substack{i=1\ldots n_s \\ i \neq i_{max}}} p_{si}(t+1)$$

We will now discuss possible estimations of θ_{sil}. The Karush-Kuhn-Tucker (KKT) conditions [23] state that the derivative of (5) with respect to t_{si} evaluated at the optimum should necessarily be zero. This means that at optimality:

$$\frac{\partial L}{\partial t_{s_0 i_0}} = -d_{s_0} U_{s_0}' \left(\sum_{i=1}^{n_{s_0}} p_{s_0 i} t_{s_0 i} \right) p_{s_0 i_0} + \sum_{l : l \in s_0 i_0} \theta_{s_0 i_0 l} = 0$$

If path si only has one bottleneck, there would only be one nonzero $\theta_{s_0 i_0 l}$ in the addition, a fact that may be used to make a first estimation of θ_{sil}. However, the link does not know the source's mean ABW. To maintain communications between elements in the network as restricted as possible, links will assume that all the sources that use them have a mean ABW equal to their ABW. The link's estimation of θ_{sil} will then be:

$$\theta_{sil} = \begin{cases} d_{si} U' \left(c_l - \rho_l \right) & \text{if } l = \operatorname*{argmin}_{l \in si} \{ c_l - \rho_l \} \\ 0 & \text{otherwise} \end{cases} \qquad (6)$$

We have assumed, for simplicity's sake, that all sources use the same known utility function. This is the cost function we will use, thus finishing the specification of the algorithm. As we will see in Sec. 5.1, the consequences of this approximation are not significant and the algorithm yields a very good estimation of the optimum. Details on how to implement this algorithm in a real network are discussed in Sec. 6.

5   Fluid-Level Simulations

5.1   Distributed Algorithm Performance

In this section we shall present some simple examples to gain some insight into the proposed framework and to verify that the resulting traffic distribution of the distributed algorithm is not far from the actual optimum. We first present fluid-level simulations to verify its behavior in an idealized context. We have also included some packet-level simulations to analyze the effect of imprecise and delayed measurements, which are presented in the next section.

The first example we will consider is the simplest one: a single source has two possible paths to choose from. The two paths have a capacity of 3.0 and 4.0 respectively. In Fig. 2 we can see the value of p_1 (the portion of traffic routed through the path with the biggest capacity) obtained by the distributed algorithm and the actual optimum, as a function of the demand generated by the source.

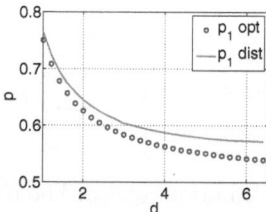

Fig. 2. Portion of traffic through the widest path in a two path single source scenario

We first remark that the distributed algorithm approximates very well the optimum, being the biggest difference less than 0.05. The second aspect that is worth noting is that the distributed algorithm always tends to "over-use" the widest path. This can be explained by the approximation we made in (6). Since $U(x)$ is concave, $U'(x)$ is a non-increasing function, meaning that if $x_1 > x_2$ then $U'(x_1) \leq U'(x_2)$. So, when a link has an ABW bigger than the source's average, its estimation of the price will be smaller than it should. In this example, it means that link 1 will calculate a smaller price, which results in the source sending more traffic through it than at optimality.

Consider now the example in Fig. 3. In it, all links have a capacity of 4.0. Source 2 generates a total demand $d_2 = 1.0$, and we analyze the optimum traffic distribution while varying d_1 (the demand generated by source 1). Notice how, even if the ABW source 1 sees on the lower path is the same as in the last example, it concentrates more traffic in the wider path than before. The presence of source 2 makes the lower path more "expensive". Also note that in this case the distributed algorithm approximates even better the global optimum.

Finally, we analyze some examples in the network of Fig. 1. In particular, we will consider the case of $c_l = 5.0 \ \forall l$ where source 1 generates a demand d_1 and the rest a demand d. The two graphs in Fig. 4 shows the optimum traffic

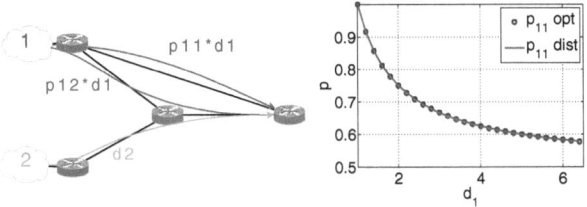

Fig. 3. The example topology, and its optimum traffic distribution as a function of d

distribution for $d = 2, 4$ as a function of d_1. The three curves represent the actual optimum, the value obtained by the distributed algorithm, and the one obtained by both MATE and TeXCP. We can see that while d_1 is relatively small and the ABW is enough, source 1 uses only the lower path. If any of these conditions is not true, p will rapidly go to 0.5, but always privileging better conditions on the upper path.

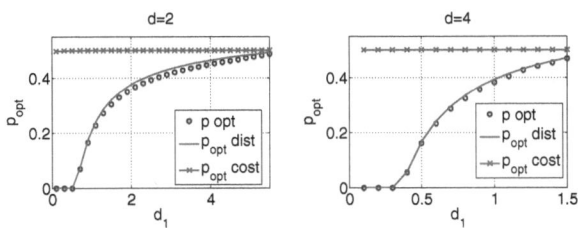

Fig. 4. Some case-scenarios for the network in Fig. 1

5.2   The Benefits of Utility Maximization

In this section we will assess the performance gain achieved by our proposal (UM from now on, as in Utility Maximization) over two well-known load-balancing techniques: MATE and TeXCP, which we already presented in Sec. 2. It is important to highlight that the results shown in this section were obtained by the distributed algorithm. We tried some centralized numerical optimization methods and obtained very similar results, verifying that approximations we did had little effect.

The comparison will be made in two real networks, with several real TMs, calculating for each of these demands two performance indicators: mean ABW perceived by sources (u_s) and link utilization (ρ_l/c_l), whose importance we have already discussed. We could consider other performance indicators, such as queueing delay or path propagation delay. However, calculation of the former depends heavily on the assumed traffic model, and we shall suppose that the latter has already been taken into account by the operator in the choice of paths.

For each TM we measured a weighted mean u_s, where the corresponding weight is $d_s / \sum d_s$. This average provides us with a rough idea of the performance as perceived by traffic. A good value of this average indicator could however hide some pathological cases where some OD pairs obtain a bad performance. That is why we also measured the 10% quantile and the minimum u_s. The comparison will be made by dividing the value obtained by UM by the one obtained by the other load-balancing technique in each case.

For each TM we also calculated the mean, 90% quantile and maximum utilization on the network for each of the mechanisms. The difference between the TeXCP indicators and the other mechanisms is presented.

Comparison in Abilene. Our first case study is Abilene [24], a well-known academic network which consists of 12 nodes and 15 bidirectional links all with the same capacity. The topology comes as an example in the TOTEM toolbox [25] and we used 388 demands (spanning a complete week) of dataset X11 from [26]. The paths we used in this case were constructed by hand, trying to give sources as much path diversity as possible, but limiting the hop count.

In Fig. 5 we can see the boxplots of the u_s indicators. We first note that the weighted mean u_s is always bigger in UM than in MATE, being generally between 1-2% and at most 4%. On the other hand, TeXCP obtains a much smaller mean u_s, generally between 4-7% and as much as 12% smaller. No conclusive results can be obtained from the quantile u_s. In the minimum u_s, UM achieves a minimum u_s that is generally between 6-12% (and can be as big as 20%) bigger than MATE. As expected, TeXCP obtains the best results in this aspect, although its gain over UM is not so large.

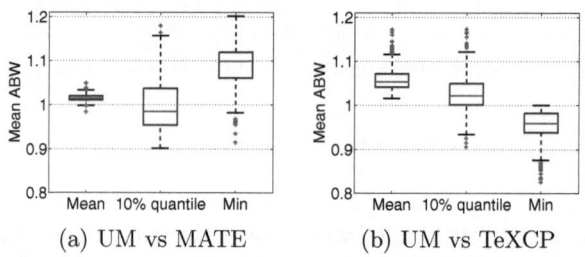

(a) UM vs MATE   (b) UM vs TeXCP

Fig. 5. u_s for UM, MATE and TeXCP in the Abilene network

Fig. 6 shows the results on link utilization. Both the mean and the quantile do not present any substantial difference between the three mechanisms (except for a relatively bigger mean utilization for TeXCP). It is in the maximum utilization that we can see a clearer distinction between them, where as expected TeXCP always obtains the best results. However, and in concordance with the u_s indicators, its gain over UM is smaller than over MATE, the former being generally 1-2% and the latter between 3-7%.

Comparison in Géant. The second case scenario is Géant [27]. This European academic network connects 23 nodes using 74 unidirectional links, with capacities that range from 155 Mbps to 10 Gbps. The topology and TMs (477 in total, covering a three week period) were obtained from TOTEM's webpage [25,5]. In

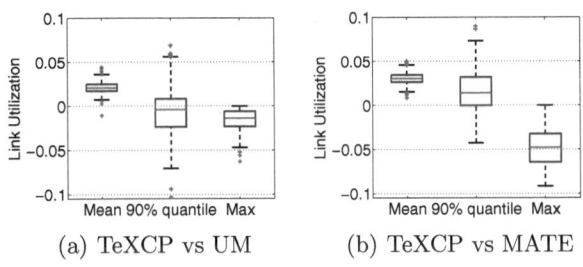

(a) TeXCP vs UM   (b) TeXCP vs MATE

Fig. 6. Link utilization for TeXCP, UM and MATE in the Abilene network

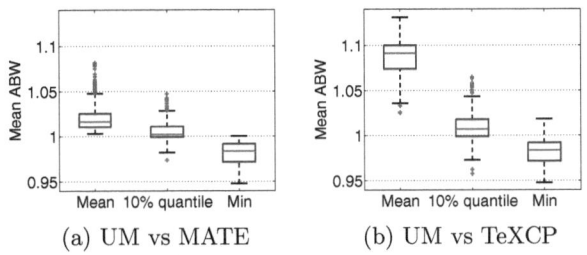

(a) UM vs MATE   (b) UM vs TeXCP

Fig. 7. u_s for UM, MATE and TeXCP in the Géant network

this case paths were constructed by a shortest path algorithm, where we used the inverse of the capacity as the link's weight. For each OD pair we computed two paths. The first is simply the shortest path, we then prune the network of the links this path uses, and compute the second shortest path.

Results for the u_s in this case can be seen in Fig. 7. This time, results of both UM and MATE are more similar, where the mean and quantile u_s are somewhat bigger for UM than MATE (although in some cases the difference easily exceeds 5%), and the minimum is relatively bigger for MATE than UM. However, the results of the comparison between UM and TeXCP are clearly in favor of the former. The mean ABW_P is generally 7-10% bigger, going as high as 13%. With respect to the minimum ABW_P, the results are logically better for TeXCP, but the difference is not significant.

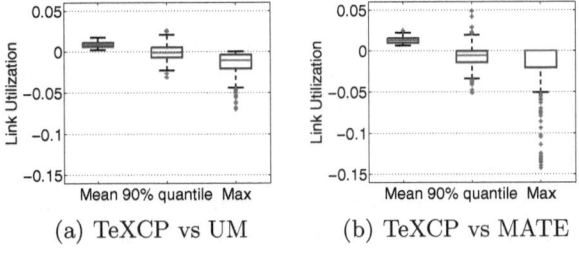

(a) TeXCP vs UM   (b) TeXCP vs MATE

Fig. 8. Link utilization for TeXCP, UM and MATE in the Géant network

Fig. 8 shows that the results for the link utilization are also very similar between UM and MATE. The difference between the two and TeXCP is not very significant, specially in the mean and quantile. With respect to the maximum utilization, TeXCP obtains only subtle improvements over the rest. However, there are some cases where the difference with MATE is more than 10%.

6   Implementation Issues

In this section we will discuss some practical issues of the distributed algorithm. Its first clear requirement is that border routers have to be able to send arbitrary portions of traffic through the different paths. Secondly, in order to measure d_{si}, interior routers should distinguish between traffic belonging to a given path that is traversing its links.

These requirements are accomplished for instance by MPLS. Hashing can be used in order to load-balance traffic with an arbitrary distribution. Packets belonging to a given si can be identified by its label header. A counter for each of them should be kept by interior routers, indicating the number of routed bytes belonging to a given label. Periodically, each router calculates the corresponding d_{si} by dividing its counter by the measurement interval, after which they reset it. In order to avoid noisy measurements, some filtering should be applied. In our simulations a simple exponential filter was sufficient. The total load ρ_l is then calculated as the sum of all d_{si} that use the link. The ABW of link l can be easily calculated as the difference between the total capacity and this value. However, in order to avoid numerical problems, the maximum between the ABW and a relatively small value (for instance $c_l/100$) should be used.

Another important aspect is the communication between link and source. Explicit messages from the source to the link are not necessary, since communication in that sense is simply how much traffic source s is sending through link l. It is true that what actually reaches the link will always be smaller or equal than originally at the source, but this approximation does not affect the algorithm's performance.

The most challenging communication is from the link towards the source. We will use the same approach as TeXCP and use probe packets, which in our case will contain the path's ABW and total cost $(\sum_{l:l \in si} \widehat{\theta}_l)$. Periodically, the source sends a probe packet, initially indicating as the path's ABW and cost ∞ and 0.0 respectively. As the probe advances towards the destination node, each interior router checks the ABW indicated in it. If this value is bigger than that of the outgoing link, the router overwrites it with the link's ABW (and "remembers" it). When the destination node receives the probe packet, he sends it back to the source through the same path but in the opposite direction. As it is going back, each interior router checks whether the final ABW indicated on the probe packet is the one its link had when the packet first passed. If this is the case, it means that he is the bottleneck of the particular path. He then calculates θ_{sil} accordingly, updates the link's total cost $\widehat{\theta}_l$, and adds this value to the total path cost indicated on the packet. Finally, the source receives the path's ABW and total cost, and updates his load distribution.

When applying the distributed algorithm, one rapidly realizes that the value of γ (the adaptation step) is very important. This value indicates how fast the probabilities adapt. A very big value makes the algorithm unstable, while a very small one makes it unresponsive. The problem is that a "good" choice of γ depends on the network topology, but also on the current load. A value that works when the network is too congested may make the network unresponsive when the network is lightly loaded. In this last case one may think that it is not very urgent to change the traffic distribution to the optimum. Research on this direction will be the object of future work.

As a final remark we will emphasize the importance of load measurement periods being smaller (several times smaller) than the inter-probe period. This way, the source can clearly appreciate the effects of the last load distribution update. If this is not the case, the distributed algorithm will either be too unstable or unresponsive.

6.1   Packet-Level Simulations

In order to verify the correct operation of the algorithm in a realistic context (i.e. delayed and unprecise measurements), we implemented it in an ns-2 [28] simulation script. The first example we will present is again the one of Fig. 1. This time, all links have a capacity of 1.0 Mbps, except for the "access" ones which have 2.0 Mbps, and all transmission delays are 20 ms. Traffic consists of elastic flows with an exponentially distributed size with mean 20 kB, arriving as a Poisson process of the corresponding intensity so as to generate a demand $d = 400$ kbps for all sources. The exponential filter's parameter is set to $\beta = 0.7$, and γ is set to 250×10^{-9} (although it may seem like a small value, we measured everything in bps). Probabilities are updated every 60 seconds and load measurements are made every 20 seconds. Fig. 9 shows the evolution of p (the upper path's portion of source 1 traffic) over time for several initial values. Two important aspects of the algorithm are highlighted in this example. First, the initial condition has virtually no effect on the stationary behavior of p. Secondly, the distributed algorithm converges to the optimum very fast (less than 15 iterations). Finally, the simulations indicate that, although they traverse a more congested link, flows on the lower path are transferred faster than those in the upper one. This can be explained by the bigger queueing delay in the upper path. The second router on the lower path has no queueing delay, because packet size is constant, and their interarrival time is shaped by the previous router. This means that, by preferring less shared links, the algorithm avoids the use of links with big queueing delays.

We will now present an example where two UM load-balancing algorithms interact. In Fig. 10 we can see a simple case scenario. There are two sources and each of them can use two paths, one of which is shared. Links as well as the traffic characteristics are the same as in the previous example. The initial probabilities are set on both sources so that the shared link is not used, maximizing the likelihood of oscillations. Probabilities are updated every 50 seconds ($\gamma = 5 \times 10^{-9}$) and the load measurements are made every 10 seconds. Sources are however not

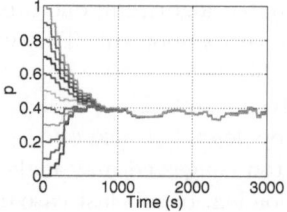

Fig. 9. Evolution of p over time in the example of Fig. 1

coordinated and update their probabilities at different moments. Both sources generate the same demand, approximately 1.1 Mbps, which the network cannot initially support. The optimal distribution is then that both sources send a third of their traffic through the shared path.

In Fig. 10 we see that both sources at first rapidly change their probability to start using the middle path. It takes them a little while to realize that another source is using this path, and start augmenting the direct path probability, but slower than before, since the price difference between them is not big now. The probabilities finally converge to the optimum after some few minutes. This whole process takes approximately 15 min. (only 20 iterations). Notice that load measurements need not be very precise, and that the algorithm supports some noise.

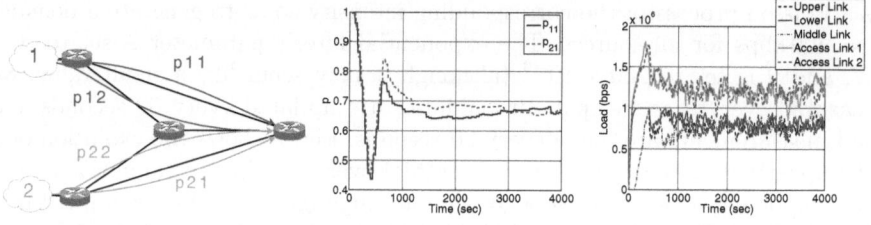

Fig. 10. The example topology, and the traffic distribution and links' load as a function of time

7 Concluding Remarks

In this work we presented a load-balancing mechanism that takes into account the needs of both the network operator and the users. We achieved this by defining an objective function that is not a cost at the *link level*, but a utility at the *OD pair level*. This lead to an optimization formulation very similar in spirit to Multi-Path TCP [15], where we maximize the sum of a utility function whose argument is the average ABW each OD pair sees in its path. Although the resulting optimization problem was not convex, a distributed algorithm was outlined which finds very tight approximations to the optimum in a relatively short time.

Along with our proposal (noted as UM), we considered two previously proposed dynamic load-balancing mechanisms: MATE [1] and TeXCP [2]. From our study, conducted over two real networks along with several real traffic demands, some conclusions can be drawn. Firstly, performance as perceived by traffic (measured as the mean ABW) is always better in UM than both MATE and TeXCP. More specifically, the improvement over MATE is generally not very big for the mean value, but can be important, specially in the worst ABW. This difference comes from the implicit unfairness among OD pairs of the social cost function of MATE. With respect to TeXCP, UM obtains a significantly better performance, specially when the link capacities are not similar. Secondly, results on link utilization are very similar for UM and TeXCP. MATE obtains similar results in the mean and quantile link utilization. However, maximum link utilization can be significantly bigger in MATE than the other two mechanisms. All in all, UM is the most balanced mechanism, in the sense that it generally outperforms the rest (though in some cases the difference may not be large), and when it does not there is only a small difference.

Much remains to be done. For instance, we have considered only elastic traffic but the performance obtained by streaming traffic should be studied. Moreover, the stability of the algorithm should be analyzed, specially considering that several approximations were used.

References

1. Elwalid, A., Jin, C., Low, S., Widjaja, I.: MATE: MPLS adaptive traffic engineering. In: INFOCOM 2001, vol. 3, pp. 1300–1309 (2001)
2. Kandula, S., Katabi, D., Davie, B., Charny, A.: Walking the tightrope: responsive yet stable traffic engineering. In: ACM SIGCOMM 2005, pp. 253–264 (2005)
3. Fischer, S., Kammenhuber, N., Feldmann, A.: Replex: dynamic traffic engineering based on wardrop routing policies. In: CoNEXT 2006, pp. 1–12 (2006)
4. Srikant, R.: The Mathematics of Internet Congestion Control, Birkhäuser Boston (2003)
5. Uhlig, S., Quoitin, B., Lepropre, J., Balon, S.: Providing public intradomain traffic matrices to the research community. SIGCOMM Comput. Commun. Rev. 36(1), 83–86 (2006)
6. Duffield, N.G., Goyal, P., Greenberg, A., Mishra, P., Ramakrishnan, K.K., van der Merwe, J.E.: Resource management with hoses: point-to-cloud services for virtual private networks. IEEE/ACM Trans. Netw. 10(5), 679–692 (2002)
7. Zhang, C., Kurose, J., Towsley, D., Ge, Z., Liu, Y.: Optimal routing with multiple traffic matrices tradeoff between average and worst case performance. In: ICNP 2005 (November 2005)
8. Ben-Ameur, W., Kerivin, H.: Routing of uncertain traffic demands. Optimization and Engineering 6(3), 283–313 (2005)
9. Juva, I.: Robust load balancing. In: IEEE GLOBECOM 2007, November 26-30, pp. 2708–2713 (2007)
10. Casas, P., Fillatre, L., Vaton, S.: Robust and Reactive Traffic Engineering for Dynamic Traffic Demands. In: NGI 2008 (April 2008)
11. Wang, W.-H., Palaniswami, M., Low, S.H.: Optimal flow control and routing in multi-path networks. Perform. Eval. 52(2-3), 119–132 (2003)

12. Kelly, F., Voice, T.: Stability of end-to-end algorithms for joint routing and rate control. SIGCOMM Comput. Commun. Rev. 35(2), 5–12 (2005)
13. Han, H., Shakkottai, S., Hollot, C.V., Srikant, R., Towsley, D.: Multi-path tcp: a joint congestion control and routing scheme to exploit path diversity in the internet. IEEE/ACM Trans. Netw. 14(6), 1260–1271 (2006)
14. Paganini, F.: Congestion control with adaptive multipath routing based on optimization. In: 40th Annual Conference on Information Sciences and Systems (March 2006)
15. Key, P., Massoulie, L., Towsley, D.: Path selection and multipath congestion control. In: IEEE INFOCOM 2007, May 2007, pp. 143–151 (2007)
16. He, J., Bresler, M., Chiang, M., Rexford, J.: Towards robust multi-layer traffic engineering: Optimization of congestion control and routing. IEEE Journal on Selected Areas in Communications 25(5), 868–880 (2007)
17. Bonald, T., Massoulié, L.: Impact of fairness on internet performance. ACM SIGMETRICS 2001, 82–91 (2001)
18. Mo, J., Walrand, J.: Fair end-to-end window-based congestion control. IEEE/ACM Trans. Netw. 8(5), 556–567 (2000)
19. Kelly, F., Maulloo, A., Tan, D.: Rate control in communication networks: shadow prices, proportional fairness and stability. Journal of the Operational Research Society 49 (1998)
20. Bonald, T., Massoulié, L., Proutière, A., Virtamo, J.: A queueing analysis of max-min fairness, proportional fairness and balanced fairness. Queueing Syst. Theory Appl. 53(1-2), 65–84 (2006)
21. Bonald, T., Proutière, A.: On performance bounds for balanced fairness. Perform. Eval. 55(1-2), 25–50 (2004)
22. Pappalardo, M.: On the duality gap in nonconvex optimization. Math. Oper. Res. 11(1) (1986)
23. Minoux, M.: Programmation Mathématique: théorie et algorithmes, Dunod (1983)
24. The Abilene Network, http://www.internet2.edu/network/
25. TOTEM: TOolbox for Traffic Engineering Methods, http://totem.info.ucl.ac.be/
26. Zhang, Y.: Abilene Dataset, http://www.cs.utexas.edu/~yzhang/research/AbileneTM/
27. Géant Topology Map, http://www.geant.net
28. The Network Simulator - ns, http://nsnam.isi.edu/nsnam/index.php/Main_Page

The Impact of Congestion Control Mechanisms on Network Performance after Failure in Flow-Aware Networks

Jerzy Domżał, Robert Wójcik, and Andrzej Jajszczyk

Department of Telecommunications,
AGH University of Science and Technology, Kraków, Poland
{jdomzal,robert.wojcik,jajszczyk}@kt.agh.edu.pl

Abstract. This paper presents the impact of congestion control mechanisms proposed for Flow-Aware Networks on packet transmission in the overloaded network after a link failure. The results of simulation based analysis show how to set the values of the congestion control parameters in order to decrease the acceptance time of the interrupted streaming flows in the backup link. The research was performed for three congestion control mechanisms, the Enhanced Flushing Mechanism (EFM), the Remove Active Elastic Flows (RAEF), and Remove and Block Active Elastic Flows (RBAEF) in two different cross-protect router architectures, with the PFQ (Priority Fair Queuing) and with the PDRR (Priority Deficit Round Robin) scheduling algorithms. Moreover, the advantages and weaknesses of using the proposed solutions in FAN, considering the effects of a network element failure, are described and analyzed.

Keywords: congestion control, Flow-Aware Networks, protection, restoration.

1   Introduction

The Quality of Service for packet networks is still a very important and interesting issue. The dominating QoS architecture in the current networks is DiffServ (Differentiated Services). Unfortunately, this architecture is complicated and in many cases does not work as expected. Currently, high quality data transmission is possible only because of overprovisioned network links. The more and more popular applications like VoIP or VoD need low delays, low packet loss and link capacities high enough to work satisfactorily. In fact, all new applications and services in packet networks require a controlled quality of connections. This necessity has triggered many studies on providing new possibilities of ensuring the proper Quality of Service in the packet networks. Flow-Aware Networking (FAN) is a new concept for packet switched networks with QoS guaranties. The main assumption of FAN is to provide maximum possible benefits in the perceived QoS using only the minimal knowledge of the network. In this paper, we argue that by using proper congestion control mechanisms in FAN, we can enhance the perceived QoS, especially in case of network failures.

R. Valadas and P. Salvador (Eds.): FITraMEn 2008, LNCS 5464, pp. 53–67, 2009.
© Springer-Verlag Berlin Heidelberg 2009

The remainder of the document is organized as follows. Section 2 introduces the general idea of FAN. Section 3 shows the congestion control mechanisms for FAN and their brief description. In Section 4, the results of carefully selected simulation experiments for each congestion control algorithm and two different queuing disciplines are presented. Section 5 concludes the paper.

2   Flow-Aware Networks

The concept of Flow-Aware Networking as a novel approach to assure quality of service in packet networks was introduced in 2004 [1]. The goal of FAN is to enhance the current IP network by improving its performance under heavy congestion. To achieve that, certain traffic management mechanisms to control link sharing are introduced, namely: measurement-based admission control [2] and priority scheduling [1], [3]. The former is used to keep the flow rates sufficiently high, to provide a minimal level of performance for each flow in case of overload. The latter realizes fair sharing of link bandwidth, while ensuring negligible packet latency for flows emitting at lower rates.

In FAN, admission control and service differentiation are implicit. There is no need for a priori traffic specification, as well as there is no class of service distinction. However, streaming and elastic flows are implicitly identified inside the FAN network. This classification is based solely on the current flow peak rate. All flows emitting at lower rates than the current fair rate are referred to as streaming flows, and packets of those flows are prioritized. The remaining flows are referred to as elastic flows. The distinctive advantage of FAN is that both streaming and elastic flows achieve a necessary quality of service without any mutual detrimental effect.

2.1   The Cross-Protect Mechanism

FAN is supposed to be an enhancement of the existing IP network. In order to function properly, an upgrade of current IP routers is required. Figure 1 shows a concept diagram of a Cross-Protect router (XP router), the interconnecting device in FAN networks. FAN adds only two blocks to the standard IP router. They are namely: admission control block and scheduling block. The former is placed in the incoming line cards of the router, whereas the latter is situated in the outgoing line cards.

Admission control is responsible for accepting or rejecting the incoming packets, based on the current congestion status. If a packet is allowed, the identifier (ID) of flow associated with it may be added to the protected flow list (PFL), and then all forthcoming packets of this flow will be accepted. The packets of new flows may be accepted in the admission control block only when the links are not congested. The ID is removed from the PFL after a specified time period of flow inactivity given by the value of the *pfl_flow_timeout* parameter. The admission control block realizes the measurement based admission control (MBAC) functionality [4]. The scheduler is responsible for queue management and it has to ascertain that all flows are equally treated.

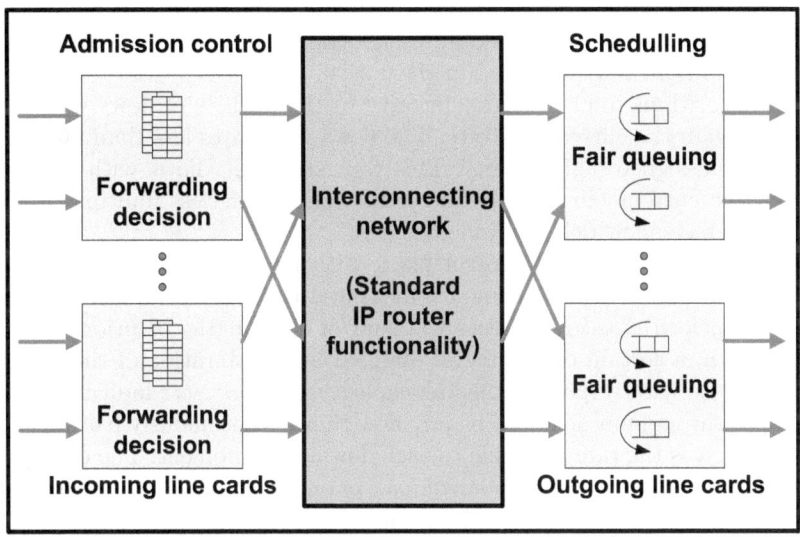

Fig. 1. Concept diagram of a Cross-Protect router [1]

Naming FAN devices as "Cross-Protect routers" is a result of a mutual cooperation and protection, which exists between both discussed blocks. The admission control block limits the number of active flows in the XP router, which essentially improves the queuing algorithm functionality, and reduces its performance requirements. It is vital that queuing mechanisms operate quickly, as for extremely high speed links the available processing time is strictly limited. On the other hand, the scheduling block provides admission control with the information on congestion status on the outgoing interfaces. The information is derived based on, for example, current queues occupancy. The cross-protection contributes to a shorter required flow list and queue sizes, which significantly improves FAN scalability.

2.2 Scheduling

A queuing algorithm, implemented in the scheduler block is the most important mechanism when considering the congestion control issue. It decides (by measurements) which flows may be served and which should be blocked. It allows for fair access to the resources without any intervention form the user. Over time, numerous queuing algorithms have been proposed [5]. In our analysis, we used FAN links with the implemented functionality of the PFQ (Priority Fair Queuing) and PDRR (Priority Deficit Round Robin) fair queuing algorithms.

PFQ is a modified version of the SFQ (Start-time Fair Queuing) algorithm [6]. PFQ inherits the advantages of SFQ and is enriched by the packet prioritizing possibilities in the scheduler module. Similarly, PDRR is an enhanced version of DRR (Deficit Round Robin) [7]. These algorithms operate differently, however, the outcome of their functioning is almost identical. A more detailed description of both queuing disciplines, including pseudocodes, measured indicators and all required definitions may be found in [1] and [3].

These algorithms implicitly give priority to the packets of flows whose peak rate is less than the current fair rate. The flows with rates less than the current fair rate are assigned high priority. This way, streaming flows with peak rates less than the current fair rate are subjects to the bufferless multiplexing and, therefore, perceive low delays and losses [8].

To provide the admission control block with proper congestion status, *priority_load* and *fair_rate* indicators are measured periodically by the scheduling block. The *priority_load* represents the sum of the lengths of priority packets transmitted in a certain time interval, divided by the duration of that interval, and normalized with respect to the link capacity. The *fair_rate* indicates approximately the throughput achieved by any flow that is continuously backlogged. In other words, it is the rate available to each flow at the moment. The detailed description of both *fair_rate* and *priority_load* parameters, along with the methods of estimating them are presented in [1] and [3].

Both congestion indicators are calculated periodically. As mentioned before, the *fair_rate* is used to differentiate between streaming and elastic flows within the XP router. Additionally, along with the *priority_load*, it is used by the admission control to selectively block new incoming flows, provided that the congestion state is detected.

3   Congestion Control Mechanisms for FAN

The congestion control mechanisms for FAN, presented in [9] and [10], were proposed to decrease the access time to the congested link for new streaming flows that may represent, for example, the VoIP connections. Users expect that after having dialed the number, their connection will go through almost immediately. Therefore, it is necessary to provide short acceptance times for such flows. The acceptable time for international calls should be not greater that 11 seconds for the 95% of the calls, while for the local calls it should not exceed 6 seconds [11]. In the basic version of the admission control algorithm, the new flows cannot be accepted in the XP routers under the state of congestion. It may cause that, in some cases, the whole bandwidth may be occupied by a finite number of flows giving no chance for new flows to begin their transmission for a long time. The new mechanisms based on the whole or partial cleaning the PFL content in the congestion state allow for decreasing the acceptance time of new streaming flows. It is possible to choose such values of the parameters characteristic for the proposed solutions that result in achieving a short acceptance time of new streaming flows, but not significantly increasing the value of the mean transmission time of elastic flows. Three versions of congestion control mechanisms; EFM, RAEF, and RBAEF are proposed and described below in details. The main goal of this paper, however, is to show that these mechanisms also work very well when a network device or a link fails and the traffic redirection is needed.

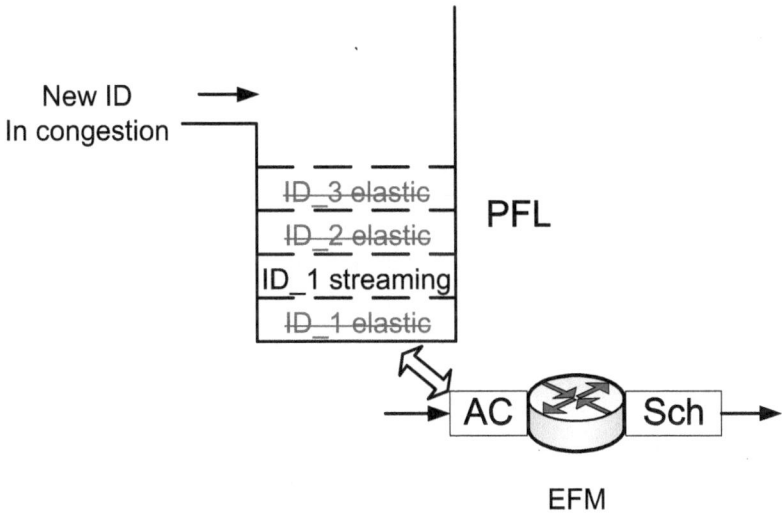

Fig. 2. The operation principle of EFM

3.1   The Enhanced Flushing Mechanism

The EFM (Enhanced Flushing Mechanism), presented in [9] and [10], is a good proposition to solve the problem of too long acceptance times of new streaming flows in the AC block. In this algorithm the identifiers of all elastic flows are removed from the PFL if a packet of a new flow comes to the AC block in the congestion state (see Figure 2). However, the PFL should not be flushed (erased) in all such cases. It is necessary to provide a time variable denoted by us as *pfl_flushing_timer*. The value of this variable represents the minimum period of time that has to expire before the next flushing action can occur. It ensures a stable operation of the algorithm. The flows which identifiers were removed from the PFL are not blocked in the AC block and can resume the transmission promptly. After removing the identifiers of the elastic flows from the PFL, the flows have to compete with each other for acceptance in the AC block and it may take some time before they will be accepted again.

Based on the results obtained in the experiment described in [10] we can conclude that it is possible to ensure short acceptance times of new streaming flows in the AC block independently of the number of elastic flows being active in the background and the number of streaming flows, which want to begin the transmission. The EFM works sufficiently well with both the PFQ and the PDRR algorithms.

The implementation of the EFM in the cross-protect router is quite simple and does not increase the complexity and power resources significantly.

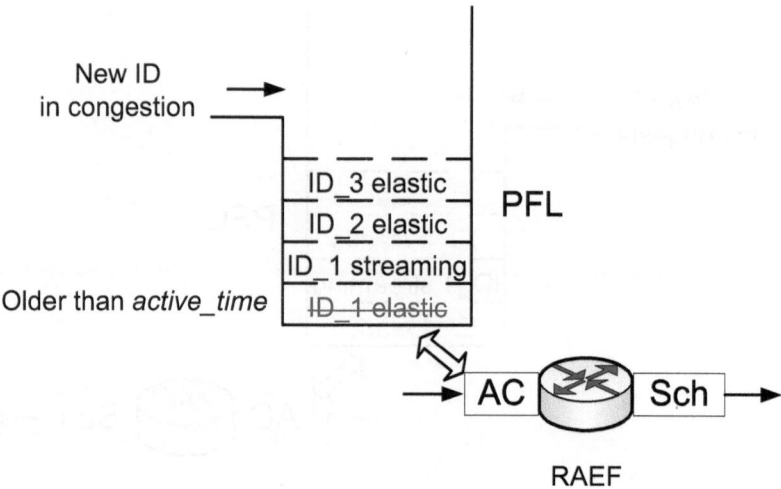

Fig. 3. The operation principle of RAEF

3.2   The RAEF Mechanism

The second mechanism proposed to solve the problem of the too long acceptance time of new streaming flows in the AC block is RAEF (Remove Active Elastic Flows). In this algorithm, only the identifiers of those elastic flows that were active for at least a specified period of time (*active_time*) are removed from the PFL when congestion is noticed (see Figure 3). The flows which identifiers were removed from the PFL are not blocked in the AC block (as in the EFM) and can resume the transmission promptly. The disadvantages of this algorithm are also the same as in the EFM. It is possible that the identifiers of such flows will not be added to the PFL again immediately or even in a short time and the transmission time of their traffic may be extended. The flows which identifiers were removed from the PFL, have to compete with other flows for acceptance in the AC block and it may take some time before such flows will be accepted again.

The results of carefully selected simulation experiments for analyzing the RAEF mechanism are presented in [9]. They show that the algorithm ensures quick acceptance times of new streaming flows in the AC block independently of the number of elastic flows being active in the background and the number of streaming flows which want to begin the transmission. Similarly to EFM, the RAEF mechanism works satisfactorily with both analyzed versions of the scheduling algorithm, the PFQ and the PDRR.

Similarly as in case of EFM, the implementation of the RAEF mechanism in the cross-protect router is simple and does not increase the complexity and power resources significantly.

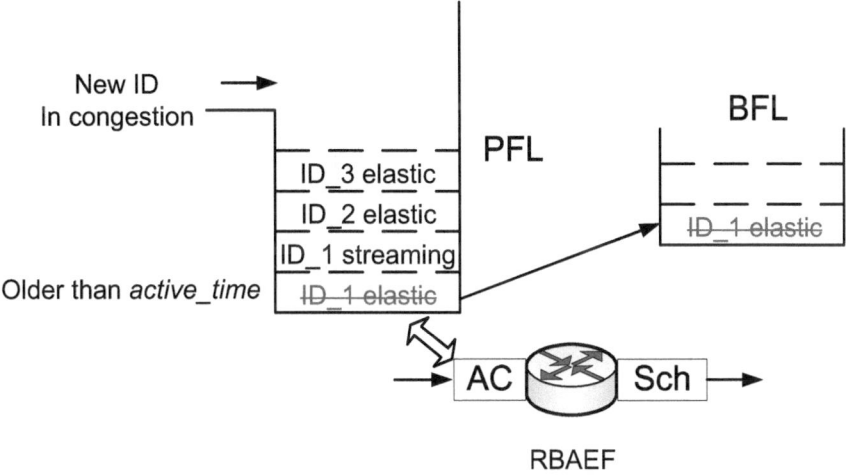

Fig. 4. The operation principle of RBAEF

3.3   The RBAEF Mechanism

The last mechanism proposed in [9] to decrease the time interval between beginning of sending the packets by a new streaming flow and its acceptance in the AC block is called RBAEF (Remove and Block Active Elastic Flows). In this algorithm the identifiers of the elastic flows being active for a specified period of time are removed from the PFL every time when congestion is noticed — just as in case of RAEF. However, the identifiers of such flows are then written to the BFL (Blocked Flow List) for a short, fixed period of time called *blocked_time* (see Figure 4). If a packet arriving to the admission control block belongs to the flow, the identifier of which is in the BFL, the packet is always dropped. Therefore, the flows removed from the PFL list can continue the transmission only after their tag has been removed from the BFL. The flows which identifiers were removed from the BFL, can continue transmission, but again, they have to compete with other flows for link resources and it may take some time before such flows will be accepted again.

The simulation scenario for analyzing the RBAEF mechanism is described in [9]. Based on the results obtained in the experiment we can conclude that this algorithm also ensures quick acceptance of new streaming flows in the AC block independently of the number of elastic flows being active in the background and the number of streaming flows which want to begin the transmission. As both previously presented mechanisms, the RBAEF mechanism also works satisfactorily with both analyzed versions of the scheduling algorithm, the PFQ and the PDRR.

The implementation of the RBAEF mechanism in the cross-protect router is slightly more complicated than in the previous cases, but does not increase the complexity and power resources significantly, either.

Table 1. Time interval between beginning of sending the packets and the acceptance of a streaming flow in the AC block in an XP router before and after L2 link failure

Mechanism	PFQ		PDRR	
	waiting time router R2 [s]	waiting time router R4 [s]	waiting time router R2 [s]	waiting time router R4 [s]
Basic FAN	78.43 ± 6.56	240.86 ± 26.64	89.28 ± 7.87	258.34 ± 30.65
EFM	0.46 ± 0.10	0.67 ± 0.22	0.98 ± 0.38	0.54 ± 0.11
RAEF	$0.37 \pm .023$	0.45 ± 0.13	0.15 ± 0.10	0.82 ± 0.67
RBAEF	0.56 ± 0.21	0.90 ± 0.23	0.05 ± 0.01	0.08 ± 0.06

4 FAN in Case of Failure

The analysis of traffic in the network is very important, especially in the case of failures. In the basic FAN architecture, the ID of a flow can be removed from the PFL list only in case of its long enough inactivity. It means that the transmission of all flows accepted in the AC block cannot be stopped and it is true until there are no failures in the network.

In this section we analyze the mean acceptance time of a new streaming flow in the topology presented in Figure 5. There is one source node and two destination nodes in our experiments. We assumed that bottleneck links L3 and L5 are FAN links of the 100 Mbit/s capacity. The capacity of the rest of the links, with the FIFO queue, was set to 1 Gbit/s. The shortest path routing was implemented in this network, which means that under normal conditions the traffic to node D1 is sent through nodes R1, R2 and R3 while the traffic to node D2 is sent through nodes R1, R4 and R5. By using such a topology we decided that link L5 is treated as a backup for the traffic sent normally through the link L3. We analyzed the effects of failures of link L2 at a chosen time instant.

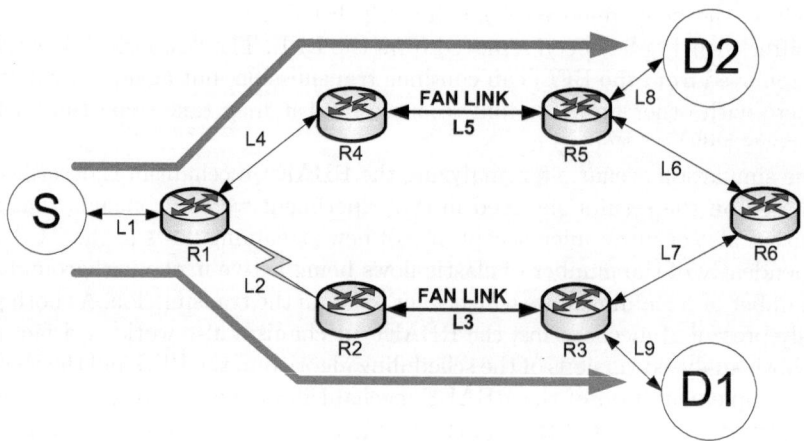

Fig. 5. Simulated network topology

We provided the traffic pattern with the Pareto distribution for calculating the volume of the elastic traffic directed to both destination nodes. The exponential distribution for generating the time intervals between beginnings of the transmissions of the elastic flows was used. The exponential distribution was also used to generate the start times of 20 streaming flows. The packet size (100 bytes) and the transmission rate (80 kbit/s) used for the streaming flows are the typical values of the VoIP stream transmission, e.g., in Skype. We made our simulation runs in various conditions changing the number of elastic and streaming flows. We analyzed the acceptance time of each streaming flow in the AC block of node R2 (before failure) and node R4 (after failure). The measurement interval for the *priority_load* parameter was set to 50 ms while the *fair_rate* values were estimated every 500 ms. The *max_priority_load* (maximum allowed value of the priority load) and the *min_fair_rate* (minimum allowed value of the fair rate) parameters were set to 70% and 5% of the link capacity, respectively, and the *pfl_flow_timeout* parameter was set to 20 s. 95% confidence intervals were calculated by using the Student's t-distribution.

The simulations were performed under various conditions. Firstly, we decided to check the mean acceptance time of new streaming flows in FAN links with the number of background elastic flows ranging from 200 to 600. The duration of each simulation run was set to 500 s. At 250 s, link L2 was turned off and the packets of all flows were sent through link L5. The redirected streaming flows had to compete for access to the L5 link along with all other elastic flows.

Basic FAN links have unacceptable values of the *waiting_time* (see Table I). It takes tens of seconds (and even hundreds of seconds for the redirected flows) before a new flow is accepted in the router. If we imagine that this flow exemplifies a VoIP call, it is obvious that the break in transmission of such a flow is much too long. Our simulations show that the acceptance time of streaming flows can be decreased by using the congestion control mechanisms presented in [9] on both examined FAN links. The comparison of the *waiting_time* values for the basic FAN with PFQ algorithm and for its modified versions with EFM (*pfl_flushing_timer* = 5 s), RAEF (*active_time* = 5 s) and RBAEF (*active_time* = 5 s, *blocked_time* = 1 s) are presented in Figure 6. We can see that difference between the values for basic FAN and its modified versions are significant. If we use the mentioned above congestion control mechanisms the new streaming flows may be accepted in the routers after less then one second in both the basic and backup links.

The presented results show that the acceptance time of new streaming flows do not depend on the number of elastic flows in the background. The simulation results for the case with 200 elastic flows are presented in Table I. In basic FAN, for both queuing algorithms, a new streaming flow is accepted in the AC block of node R2 after tens of seconds while after L2 link failure this time (observed at node R4) raises to a few hundreds seconds. The *waiting_time* period in the R4 router represents the amount of time in which the streaming flow struggles for acquiring the backup link's resources. The difference between acceptance time of a new streaming flow in routers R2 and R4 is significant. A new flow may be accepted in router R2 when one or more elastic flows finish their transmission

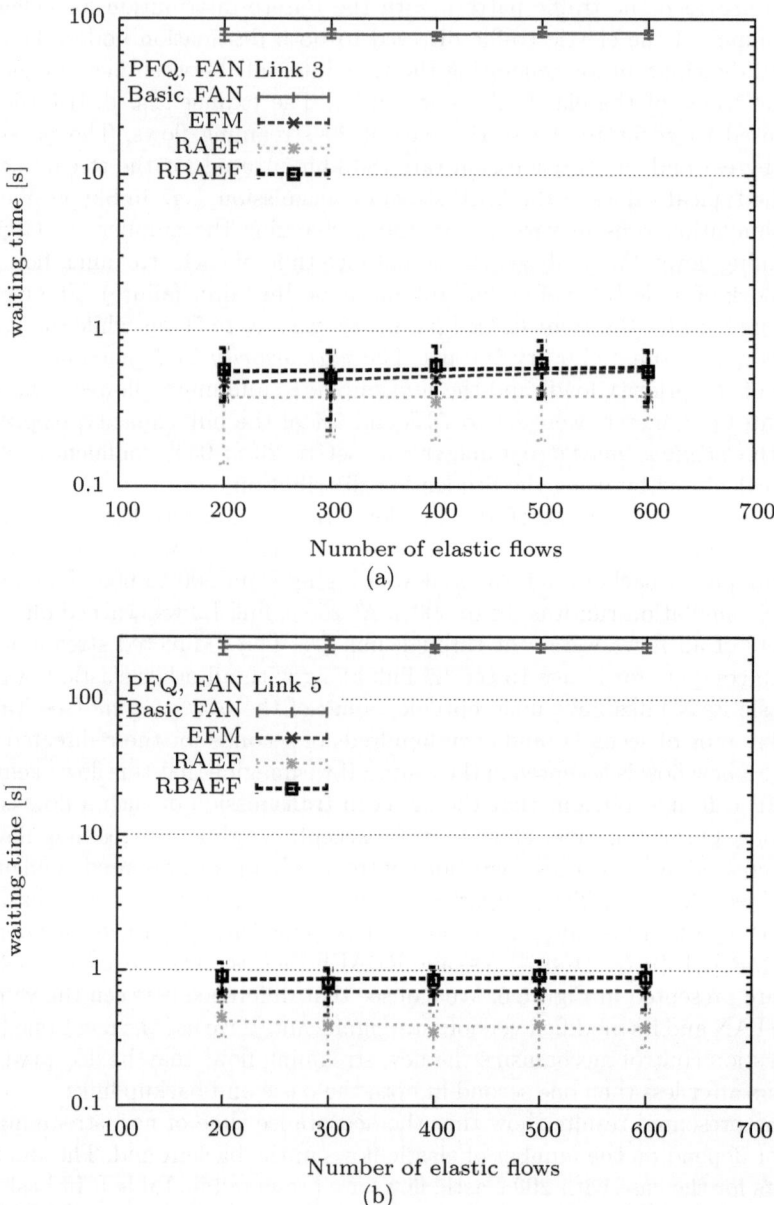

Fig. 6. The mean acceptance time of new streaming flows in basic FAN and its versions with EFM, RAEF and RBAEF, (a) for Link L3, (b) for Link L5

which allows for increasing the *fair_rate* values. After a failure, all redirected flows have to compete for the access to the R4 router which significantly increases the number of competitors in that node. Successively, this decreases their chance for being accepted and, therefore, increases the mean acceptance time.

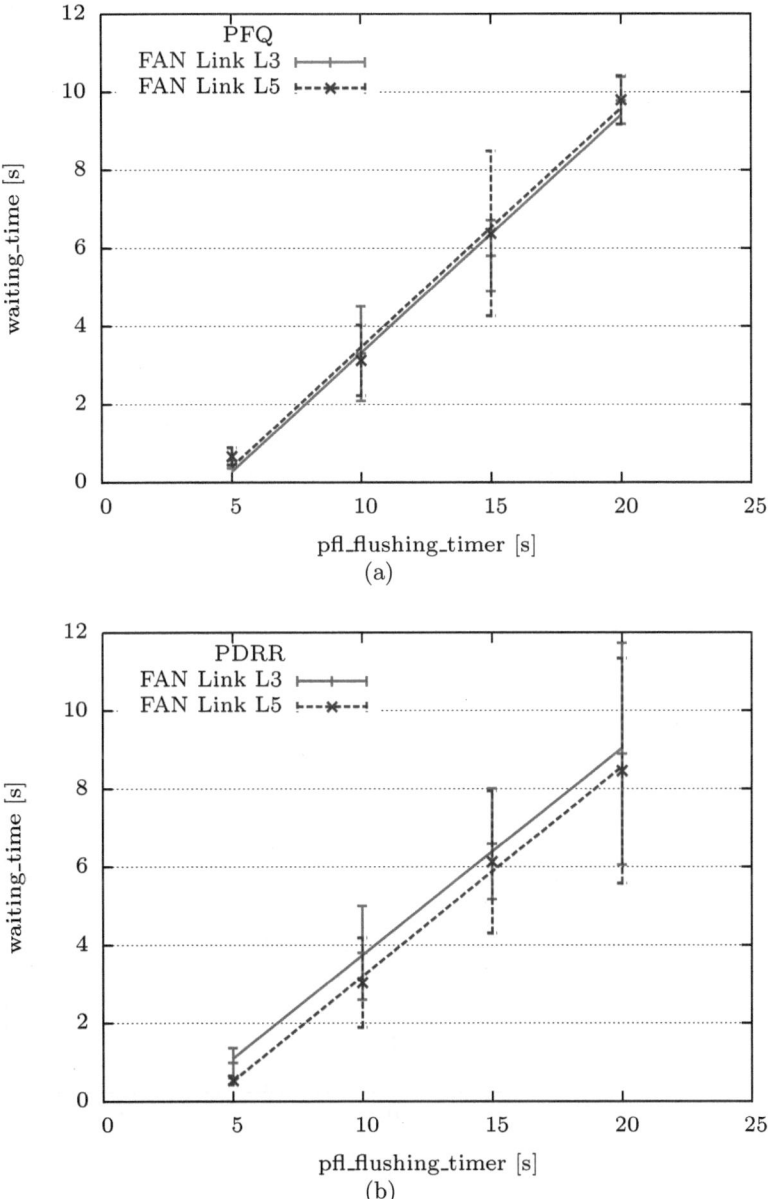

Fig. 7. The mean acceptance time of new streaming flows in the FAN nodes with EFM, (a) with PFQ scheduling algorithm, (b) with PDRR scheduling algorithm

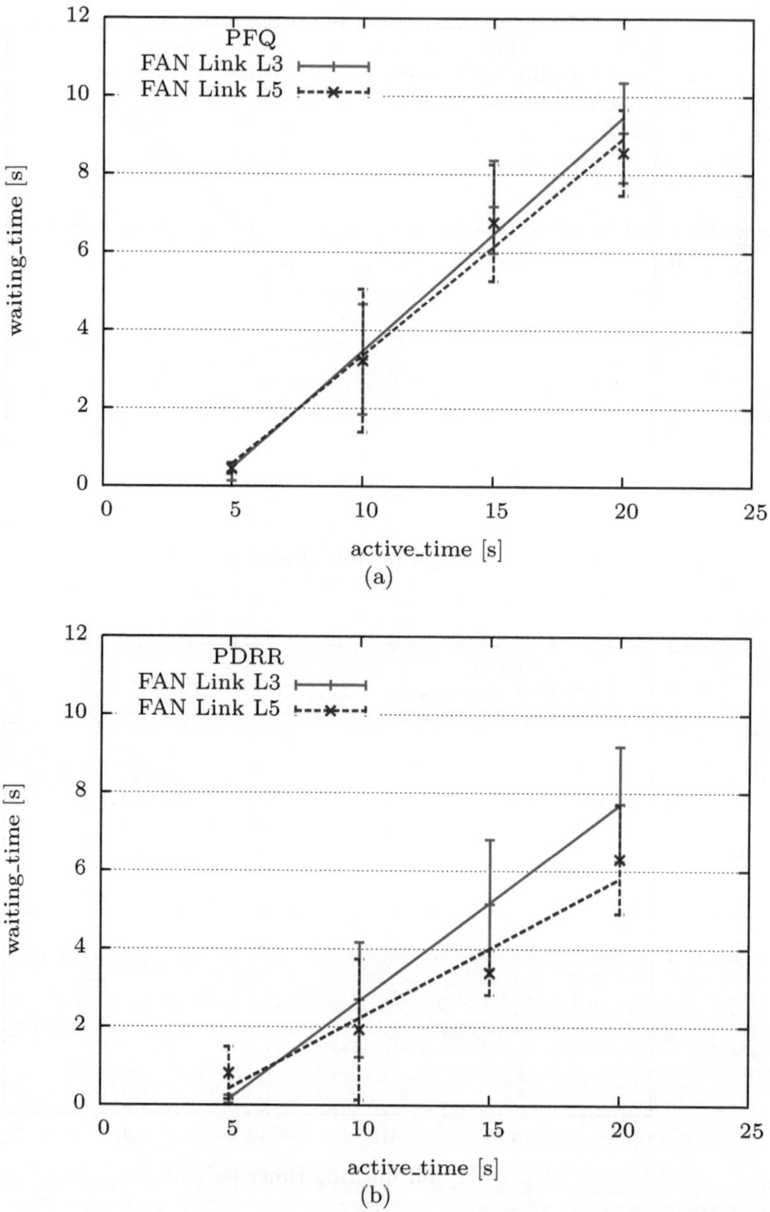

Fig. 8. The mean acceptance time of new streaming flows in the FAN nodes with RAEF, (a) with PFQ scheduling algorithm, (b) with PDRR scheduling algorithm

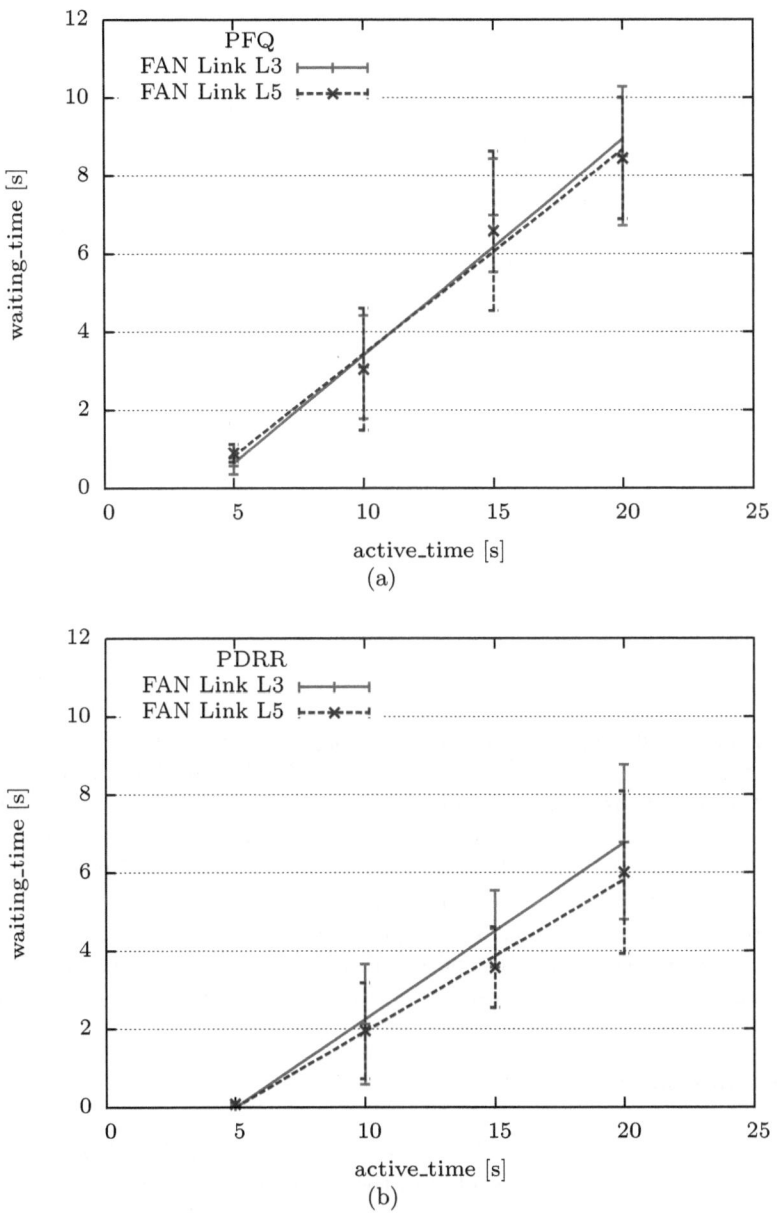

Fig. 9. The mean acceptance time of new streaming flows in the FAN nodes with RBAEF, (a) with PFQ scheduling algorithm, (b) with PDRR scheduling algorithm

The mean *waiting_time* values for all three congestion control mechanism and both queuing disciplines are presented in Table 1 and in Figures 7, 8 and 9. The values of *pfl_flushing_timer* and *active_time* parameters were set to 5 s, while the *blocked_time* threshold in RBAEF was set to 1 s.

In the EFM mechanism (Figure 7) we present the mean waiting time with respect to the *pfl_flushing_timer* parameter. This parameter defines the minimum time period that has to expire before a new flushing can occur. Under normal circumstances, a new streaming flow has to wait for the congestion to end, however, flushing mechanism allows for faster acceptance of such flows. By using EFM, a flow must only wait for the nearest flushing procedure. Therefore, a statistic streaming flow will only need a half of the *pfl_flushing_timer* to be accepted on a link. This explains the linear growth of the dependency presented in Figure 7.

Figures 8 and 9 present similar dependencies but in case of these flushing mechanisms, the *pfl_flushing_timer* is replaced by the *active_time* parameter. Now, the PFL may be flushed each time a new flow requires it, however, only the flows that were active longer than the *active_time* are erased. By analogy, we can observe that, statistically, a new streaming flow must wait for no longer than a half of the *active_time*. Moreover, the mentioned dependencies are similar for all flushing mechanisms and for both queuing disciplines.

Based on the obtained results, we may conclude that the congestion control mechanisms proposed for FAN significantly improve the performance of traffic classified as streaming. It is possible to set the values of the *active_time* (for RAEF and RBAEF) and the *pfl_flushing_timer* (for EFM) parameters so that they ensure very short acceptance time of streaming flows in a backup link after a failure. It is a strong advantage of these congestion control mechanisms. There are, however, some drawbacks of the proposed solution. The most important one is that the transmission of some elastic flows may be ceased by each flushing procedure. Therefore, it may increase the total transmission time of those flows. However, in [9], it is shown that the mean total transmission time of the elastic flows does not necessarily increase by using these congestion control mechanisms. It is possible to set the transmission parameters to such values, that allow for a significant decrease of the acceptance times of all the streaming flows without increasing the mean total transmission time of the elastic flows significantly.

5   Conclusion

The traffic management in FAN is described in this paper. In the congestion-less state, a new flow is always accepted in the AC block of an XP router while, under congestion, no new IDs can be added to the PFL. Three congestion control mechanisms, namely: EFM, RAEF and RBAEF that allow for fast acceptance of new streaming flows in the FAN router are briefly introduced. By analyzing the traffic in a simple FAN topology it is shown that these congestion control mechanisms significantly decrease the waiting time for all streaming flows. Moreover, they allow for fast acceptance of the streaming flows in the backup link after primary link's failure, which enhances FAN's network restoration capabilities. It is important to note that the backup link may also be used for normal transmission in the failure-less state and, after a primary link's failure, streaming flows from that link can be transferred via the backup link with hardly any interruptions. The IDs of the elastic flows in the backup link are removed from the PFL and have to compete for access to the resources with all flows that want to send the packets, including the flows from the broken primary link.

The solutions proposed in the paper are flexible, stable and give the chance for more reliable transmission in Flow-Aware Networks.

Acknowledgement

This work was done within the EU FP7 NoE Euro-NF (http://www.euronf.org) framework. The reported work was also supported by the Foundation for Polish Science and the Polish Ministry of Science and Higher Education under grant N517 013 32/2131.

References

1. Kortebi, A., Oueslati, S., Roberts, J.: Cross-protect: implicit service differentiation and admission control. In: IEEE HPSR 2004, Phoenix, USA (April 2004)
2. Oueslati, S., Roberts, J.: A new direction for quality of service: Flow-aware networking. In: NGI 2005, Rome, Italy (April 2005)
3. Kortebi, A., Oueslati, S., Roberts, J.: Implicit Service Differentiation using Deficit Round Robin. In: ITC19, Beijing, China (August/September 2005)
4. Kortebi, A., Oueslati, S., Roberts, J.: MBAC algorithms for streaming flows in Cross-protect. In: EuroNGI Workshop, Lund, Sweden (June 2004)
5. Suter, B., Lakshman, T.V., Stiliadis, D., Choudhury, A.K.: Buffer management schemes for supporting TCP in gigabit routers with per flow queuing. IEEE Journal on Selected Areas in Communications 17, 1159–1169 (1999)
6. Goyal, P., Vin, H.M., Cheng, H.: Start-time Fair Queuing: A Scheduling Algorithm for Integrated Services Packet Switching Networks. IEEE/ACM Transactions on Networking 5, 690–704 (1997)
7. Shreedhar, M., Varghese, G.: Efficient Fair Queuing Using Deficit Round-Robin. IEEE/ACM Transactions on Networking 4, 375–385 (1996)
8. Roberts, J.: Internet Traffic, QoS and Pricing. Proceedings of the IEEE 92, 1389–1399 (2004)
9. Domzal, J., Jajszczyk, A.: New Congestion Control Mechanisms for Flow-Aware Networks. In: IEEE ICC, Beijing, China (May 2008)
10. Domzal, J., Jajszczyk, A.: The Flushing Mechanism for MBAC in Flow-Aware Networks. In: NGI 2008, Krakow, Poland (April 2008)
11. ITU-T: Network grade of service parameters and target values for circuit-switched services in the evolving ISDN. Recommendation ITU-T E.721 (May 1999)

On the Dependencies between Internet Flow Characteristics

M. Rosário de Oliveira[1], António Pacheco[1], Cláudia Pascoal[1], Rui Valadas[2], and Paulo Salvador[2]

[1] Universidade Técnica de Lisboa
Instituto Superior Técnico and CEMAT
Av. Rovisco Pais, 1049-001 Lisboa, Portugal
{rsilva,apacheco}@math.ist.utl.pt,
151285@isabelle.math.ist.utl.pt
[2] Universidade de Aveiro
Instituto de Telecomunicações
Campus Universitário de Santiago, 3810-193 Aveiro, Portugal
{rv,salvador}@ua.pt

Abstract. The development of realistic Internet traffic models of applications and services calls for a good understanding of the nature of Internet flows, which can be affected by many factors. Especially relevant among these are the limitations imposed by link capacities and router algorithms that control bandwidth on a per-flow basis. In the paper, we perform a statistical analysis of an Internet traffic trace that specifically takes into account the upper-bounds on the duration and rate of measured flows. In particular, we propose a new model for studying the dependencies between the logarithm of the size, the logarithm of the duration, and the logarithm of the transmission rate of an Internet flow. We consider a bivariate lognormal distribution for the flow size and flow duration, and derive estimators for the mean, the variance and the correlation, based on a truncated domain that reflects the upper-bounds on the duration and rate of measured flows. Moreover, we obtain regression equations that describe the expected value of one characteristic (size, duration or rate) given other (size or duration), thus providing further insights on the dependencies between Internet flow characteristics. In particular, for flows with large sizes we are able to predict durations and rates that are coherent with the upper-bound on transmission rates imposed by the network.

1 Introduction

Investigating the nature of Internet flows is of fundamental importance to understand how the applications, the user behavior, and the various network mechanisms impact the Internet traffic. Comprehending the flow characteristics is important, for example, to understand the extent to which applications's performance could be improved by increased transmission rates, to design algorithms to control per-flow bandwidth at routers, and to obtain better models of Internet traffic.

Several studies have addressed the dependencies between the durations, sizes, and rates of Internet flows; see in particular [1,2,3,4,5]. These works rely on empirical

R. Valadas and P. Salvador (Eds.): FITraMEn 2008, LNCS 5464, pp. 68–80, 2009.
© Springer-Verlag Berlin Heidelberg 2009

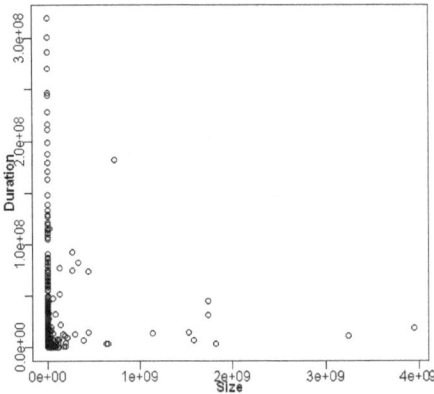

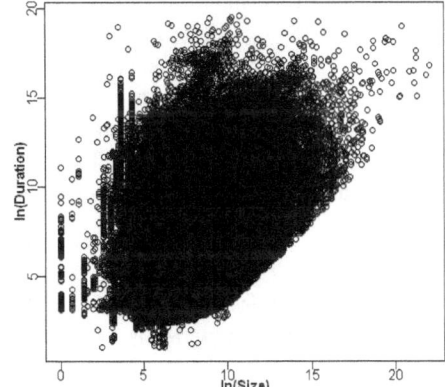

Fig. 1. Flow size (bytes) versus flow duration (ms)

Fig. 2. Logarithm of flow size versus logarithm of flow duration

estimates, such as the Pearson and Kendall's tau sample correlation coefficients, without taking into account the setting inherent to the observation of Internet flows. Indeed, the measurement of these flows is constrained by two factors. First, the observation period is finite, depending mainly on the rate of data collection and on the amount of available storage. Typical data sets considered in these studies have observation periods ranging from several minutes to several days. Second, the network topology and the supported traffic impose an upper-bound on the flow rate. In this paper, we propose a new model for studying the dependencies observed in Internet flow characteristics. We consider a bivariate lognormal distribution for the flow size and flow duration, and derive estimators for the mean, the variance, and the correlation based on a truncated domain that reflects the upper-bounds on the duration and rate of measured flows. Thus, our estimates incorporate the constraints associated with the measurement process of Internet flows. Moreover, we derive regression equations that describe the expected value of one characteristic (size, duration or rate) given other (size or duration), for both the truncated and non-truncated domains. These equations provide further insights on the dependencies between Internet flow characteristics.

To illustrate the use of our model we resort to the Bell Labs I trace, captured at the outside of the firewall connecting researchers at Bell Labs via a 9 MBits/sec link to the Internet. The network serves a total of 450 people (technical and administrative staff). Users only access HTTP servers located outside the network. The capture started at 0:00 of May 19, 2002 and ended at 16:00 of May 23, 2002, covering a total of 136 hours. The trace was processed by Tstat (http://tstat.tlc.polito.it) to detect the TCP flows from the (individual) packet data. All flows that were not totally transmitted during the capture period were filtered out by Tstat. In Figures 1 and 2 we plot the flow sizes (bytes) versus flow durations (milliseconds, ms) in linear and logarithmic scales, respectively.

The lognormal distribution has been widely used to model Internet flow characteristics [6,1,7,2,8,3,5], mainly due to its long tail. Accordingly, we assume that the joint distribution of the logarithms of Internet flow sizes and flow durations is bivariate normal. Estimating the parameters of a bivariate normal distribution is a trivial task. However, from Figure 2 it is apparent that measured data is not exactly bivariate normal but, due to the physical constraints mentioned above, is truncated on the logarithm of duration and on the logarithm of rate. Thus, it is possible to derive (improved) estimators for the parameters of the bivariate normal distribution, taking into account that measured data comes from a truncated bivariate normal distribution.

Let X denote the logarithm of flow size and Y denote the logarithm of flow duration. The truncation on the duration is represented through the condition

$$Y \leq b \tag{1}$$

where $b \geq 0$. In our study, b was set to the logarithm of the capture interval length. As the logarithm of the flow rate is $X - Y$, the truncation on the rate is represented through the condition

$$Y \geq X - k \tag{2}$$

where $k \geq 0$. In our study, k was set to the logarithm of the maximum measured flow rate. Thus, the domain of truncation is defined by

$$D = \left\{ (x, y) \in \mathbb{R}^2 : y \leq b \wedge y \geq x - k \right\} \tag{3}$$

with $k \geq 0$ and $b \geq 0$.

The paper is organized as follows: in Section 2 we make a brief description of the work related with this topic; in Section 3 we study a random pair with bivariate normal distribution truncated to D (see (3)); in Section 4 we estimate the parameters under study for the Bell Labs I trace and discuss the associations between the logarithms of flow size, flow duration, and flow rate; finally in Section 5 we draw some conclusions.

2   Related Work

The dependence between characteristics of Internet flows has been the subject of several works, including in particular [1,2,3,4,5]. Zhang et al. [5] studied 8 data sets where flows with durations smaller than a certain threshold were ignored. Based on their analysis, they have claimed that flow rates can often be described by a lognormal distribution, in agreement with [9]. Zhang et al. [5] divided the flows in four categories according to duration, in slow and fast flows, and size, in big and small flows, and investigated the correlation between the logarithm of flow rates, sizes, and durations for each of the flow categories. They have concluded in particular that the logarithm of flow sizes and the logarithm of flow rates are strongly positively correlated. Lan and Heidemann [3] did something similar to [5], including in addition rate and burstiness thresholds to divide the flows in categories, and considered several partitions of the traces. They concentrated their attention in what they called the heavy-hitter flows (flows that form a small set that carry the majority of the bytes) and evaluated the correlation between some flows characteristics, arriving at some strong correlations between some combinations of size, rate, and burstiness, which they attributed to transport and application-level protocol mechanisms. Hernandez et al. [1] used the concept of "asymptotic independence" and classical multivariate extreme value theory to analyze the relation between flow size and flow rate for large flow durations, arriving to the conclusion that for large flow durations, flow rate tends to be more closely related to flow duration, and essentially independent of flow size. Markovich and Kilpi [4] also use extreme value theory to discuss this issue, using the Pickands's A-function to estimate measures of dependency between size, duration and rate. They conclude that for extreme values, size and duration are dependent random variables, rate and size are weak dependent, and rate and duration are almost independent. Note that this last finding contradicts the work of Hernandez *et al.* [1]. The somehow conflicting results obtained in [5] and [1], suggest the use of other methodologies to address the issues in question. In particular, Kilpi *et al.* [2] used copulas for this effect.

In the paper we address the relationship between the logarithm of flow rate, the logarithm of flow duration, Y, and the logarithm of flow size, X. We let (U, V) denote the vector (X, Y) conditional to (X, Y) belonging to D, so that $W = U - V$ is the logarithm of flow rate truncated to D. A novelty of our work is the consideration of a truncated domain for the pair (U, V) which is a function of the maximum transmission rate in the network, k, and takes into account the fact that the collection periods of traces are finite.

It is possible to estimate the correlation coefficient between the unconstrained variables X and Y associated to (U, V), ρ, based on a sample of (U, V) and write the regression equations $E(U|V = y)$ and $E(V|U = x)$, for x and y belonging to the support of U and V. Moreover,

$$E(W|U = x) = x - E(V|U = x) \tag{4}$$

and

$$E(W|V = y) = E(U|V = y) - y. \tag{5}$$

Besides estimating the correlation coefficient between X and Y (which helps understanding the relationship between X and Y), the regression equations show the expected behavior of the logarithm of the rate taking into account prior knowledge on the logarithms of size or duration, and the expected behavior of the logarithm of duration when the logarithm of size is known (and *vice-versa*). The regression equations are estimated for the truncated and non-truncated model. Moreover, in order to compare these equations with those obtained by classical regression analysis, least squares linear regression estimates were also derived. Note that none of the usual assumptions considered in least squares estimation are validated, since our goal is to illustrate the differences between the sets of estimates, and not to adjust a linear regression to data.

3   Truncated Bivariate Normal Distribution

In this section we study the vector

$$(U, V) \stackrel{d}{=} (X, Y) \,|\, (X, Y) \in D \tag{6}$$

where $(X, Y) \sim N\left(\mu_1, \mu_2, \sigma_1^2, \sigma_2^2, \rho\right)$ and D is defined through (3). We first note that the joint density of (U, V), $f_{(U,V)}(x, y)$, is

$$\frac{e^{-\frac{1}{2(1-\rho^2)}\left[\frac{(x-\mu_1)^2}{\sigma_1^2} - 2\rho\frac{(x-\mu_1)(y-\mu_2)}{\sigma_1\sigma_2} + \frac{(y-\mu_2)^2}{\sigma_2^2}\right]}}{\sigma_1\,\sigma_2\,p\,2\,\pi\,\sqrt{1-\rho^2}}$$

for $(x, y) \in D$, and is 0 otherwise, where

$$p = \int_{-\infty}^{a} \phi\,(y)\,\Phi\left(\frac{\frac{y-c}{d} - \rho y}{\sqrt{1-\rho^2}}\right)\,dy \tag{7}$$

with

$$\begin{cases} a = \dfrac{b - \mu_2}{\sigma_2} \\[2mm] c = \dfrac{-k}{\sigma_2} + \dfrac{\mu_1}{\sigma_2} - \dfrac{\mu_2}{\sigma_2} \\[2mm] d = \dfrac{\sigma_1}{\sigma_2} \end{cases}.$$

We note that $\phi\,(\cdot)$ denotes the density function of the standard normal distribution and $\Phi(\cdot)$ its distribution function.

Even though not presented here, [10] has also obtained the marginal distribution of U and V and their expected values, and concluded that the marginal distributions are not truncated univariate normals.

3.1 Conditional Distributions

Conditional distributions and their expected values are an important tool for understanding the relation between two random variables. In this respect, it is convenient to note that (see [10])

$$U|V = y \sim N_{(-\infty, y+k)}\left(\mu_1 + \rho\sigma_1\frac{y - \mu_2}{\sigma_2}, \sigma_1^2\,(1 - \rho^2)\right)$$

for $y \le b$, and

$$V|U = x \sim N_{(x-k, b)}\left(\mu_2 + \rho\sigma_2\frac{x - \mu_1}{\sigma_1}, \sigma_2^2\,(1 - \rho^2)\right)$$

for $x \le b + k$, where $N_{(f,l)}\,(\mu, \sigma^2)$ denotes the univariate normal distribution with expected value μ and variance σ^2 truncated to the interval $[f, l]$.

Properties and analytical results associated with the $N_{(f,l)}\,(\mu, \sigma^2)$ distribution are presented in [11,10]. In this paper, we just present its expected value, $\mu_{(f,l)}$, because this result is used in the subsequent sections. Thus, for $f, l \in \bar{\mathbb{R}}$, where $\bar{\mathbb{R}} = \mathbb{R} \cup \{-\infty, +\infty\}$, we can write

$$\mu_{(f,l)} = \mu - \frac{\sigma}{q}\left[\phi\left(\frac{l - \mu}{\sigma}\right) - \phi\left(\frac{f - \mu}{\sigma}\right)\right],$$

where $q = \Phi\left(\frac{l-\mu}{\sigma}\right) - \Phi\left(\frac{f-\mu}{\sigma}\right)$.

As a consequence, the expected values of $U|V = y$ and $V|U = x$ are given by [10]

$$E\left(U|V = y\right) = \mu_1 + \rho\sigma_1 \frac{y - \mu_2}{\sigma_2}$$

$$- \sigma_1\sqrt{1 - \rho^2}\,\frac{\phi\left(\frac{y+k-\left(\mu_1+\rho\sigma_1\left(\frac{y-\mu_2}{\sigma_2}\right)\right)}{\sigma_1\sqrt{1-\rho^2}}\right)}{\Phi\left(\frac{y+k-\left(\mu_1+\rho\sigma_1\left(\frac{y-\mu_2}{\sigma_2}\right)\right)}{\sigma_1\sqrt{1-\rho^2}}\right)} \tag{8}$$

for $y \leq b$, and

$$E\left(V|U = x\right) = \mu_2 + \rho\sigma_2 x^* - \sigma_2\sqrt{1 - \rho^2}\times$$

$$\frac{\phi\left(\frac{b-(\mu_2+\rho\sigma_2 x^*)}{\sigma_2\sqrt{1-\rho^2}}\right) - \phi\left(\frac{(x-k)-(\mu_2+\rho\sigma_2 x^*)}{\sigma_2\sqrt{1-\rho^2}}\right)}{\Phi\left(\frac{(x-k)-(\mu_2+\rho\sigma_2 x^*)}{\sigma_2\sqrt{1-\rho^2}}\right) - \Phi\left(\frac{b-(\mu_2+\rho\sigma_2 x^*)}{\sigma_2\sqrt{1-\rho^2}}\right)} \tag{9}$$

for for $x \leq b + k$, with $x^* = (x - \mu_1)/\sigma_1$.

The expected value of the conditional variables can be written as the sum of a linear term and a non-linear term, where the linear term is equal to the conditional expected value of the conditional variables of non-truncated normal random vectors. The non-linear term shows the effect of the truncation in bivariate normal distributions. More precisely, we can say that

$$\begin{cases} E\left(U|V = y\right) = E\left(X|Y = y\right) + \zeta\left(y\right) \\ E\left(V|U = x\right) = E\left(Y|X = x\right) + \xi\left(x\right) \end{cases}$$

where $\zeta\left(y\right)$ and $\xi\left(x\right)$ denote the non-linear terms of equations (8) and (9) and x and y taking values in the supports of U and V.

Generally speaking, we can say that the conditional distributions associated with a truncated bivariate normal are the conditional distributions of the corresponding non-truncated bivariate normal, truncated to the domain under study. Thus, we can write,

$$\begin{cases} (U|V = y) \stackrel{d}{=} (X|Y = y)\,|(X,Y) \in D \\ (V|U = x) \stackrel{d}{=} (Y|X = x)\,|(X,Y) \in D \end{cases}$$

3.2 Parameter Estimation Procedure

There are several methods to estimate the unknown parameters μ_1, μ_2, σ_1^2, σ_2^2 and ρ associated with the random pair (U, V) with truncated bivariate normal distribution. We start by transforming our estimation problem into an equivalent one where we estimate the parameters of a bivariate normal distribution truncated in an unbounded rectangle with the general form: $[h, +\infty) \times [g, +\infty)$. In the new transformed space we choose the estimation methodology proposed by [12] not only because it is easy to implement, but also because it avoids the use of numerical methods to reach the solution. The transformation of the variables that will change our domain to the domain

$$D' = \{(z, w) \in \mathbb{R}^2 : z \geq h \wedge w \geq g\}$$

is

$$\begin{cases} Z' = -V \\ W' = V - U \end{cases}$$

with the identification $h = -b$ and $g = -k$.

Note that

$$(Z', W') \overset{d}{=} (X', Y') \,|\, (X', Y') \in D'$$

with $(X', Y') \sim N\left(\mu_1', \mu_2', \sigma_1'^2, \sigma_2'^2, \rho'\right)$,

$$\begin{cases} \mu_1' = -\mu_2 \\ \mu_2' = \mu_2 - \mu_1 \end{cases} \quad,$$

$$\begin{cases} \sigma_1' = \sigma_2 \\ \sigma_2' = \sqrt{\sigma_2^2 + \sigma_1^2 - 2\rho\sigma_2\sigma_1} \end{cases} \quad,$$

and

$$\rho' = \frac{\rho\sigma_1 - \sigma_2}{\sqrt{\sigma_2^2 + \sigma_1^2 - 2\rho\sigma_1\sigma_2}}.$$

After getting the estimates of μ_1', μ_2', $\sigma_1'^2$, $\sigma_2'^2$, and ρ' we need to transform them in order to get the estimates of μ_1, μ_2, σ_1^2, σ_2^2, and ρ in the following way

$$\begin{cases} \mu_1 = -(\mu_1' + \mu_2') \\ \mu_2 = -\mu_1' \end{cases} \quad,$$

$$\begin{cases} \sigma_1^2 = \sigma_1'^2 + 2\rho' \sigma_1' \sigma_2' + \sigma_2'^2 \\ \sigma_2^2 = \sigma_1'^2 \end{cases} \quad,$$

and

$$\rho = \frac{\sigma_1' + \rho' \sigma_2'}{\sigma_1}.$$

4   Results

In this section we analyze the Bell Labs I trace and its application to our model. Note that according to the notation defined in (3), $b = \ln(489\,599\,971.013) \simeq 20.009$ since the exact capture time is 135 hours 59 minutes and 59.971013 seconds (about 5 days and 16 hours), and $k = 7.194$ is the maximum observed logarithm of the rate. From Table 1 we can confirm that the maximum observed rate is 1331 Kbytes/s.

4.1   Preliminary Data Analysis

Table 1 displays some summary statistics of the trace. It is quite easy to note the large range of flow sizes, varying from 1 byte to 3.941 Gbytes. However, the analysis of the data leads to the conclusion that 95% of these observations have values under 41.75 Kbytes, leading to the conclusion that there are relatively few large observations. Even though they may be relatively few, they clearly affect the sample estimate of the expected value as the mean takes a value much larger than the sample median.

In order to allow a better understanding of the Bell Labs I data, we plot in Figure 3 the density plot [13] of the logarithm of flow size versus logarithm of flow duration. There are several regions and stripes where the densities are higher. A horizontal stripe corresponds to a predominant duration for which several sizes occur. There is a horizontal stripe at approximately $y = 13$ (10 minutes). It is caused by the Tstat timeout: all flows that do not expire according to the standard TCP timeout values are considered expired after 10 minutes without traffic. There is a thick horizontal stripe at $y \in [1.5, 9]$ (durations between 8 s and 100 s) which correspond to some typical TCP session timeout values. A region with very high density occurs at $y \simeq 6$ and $x \in [6, 12]$ (durations of around 500 ms and sizes from 1 Kbyte to 150 Kbytes). This corresponds to successful transfers of small data files, e.g. common web objects and email contents, generated by E-mail POP, download of Web page items using non-persistent HTTP connections, and control messages and the download of file chunks in P2P file sharing applications. The flows in the region defined by $y \in [2.5, 6]$ and $x \in [4.5, 7]$ (durations from 12 ms to 500 ms and sizes between 90 bytes to $\simeq 1$ Kbyte) are mainly P2P File sharing connection attempts and control messages, 304 Not Modified answers from web servers to conditional GET requests (this type of requests can be frequent and have their origin in automatic (or manual) page reloads) and 404 Not Found answers from web servers. Finally, there are two vertical stripes at $x \simeq 2$ and $x \simeq 3$ with $y \in [4, 15]$ (very small sizes, 0 - 60 bytes, and a wide range of durations, 50 ms - 60 minutes). This corresponds to streams of acknowledgments associated with uploaded files (using file sharing and FTP) and TCP port scans.

Table 1. Summary statistics of the flow size, flow duration, and flow rate, along with logarithms of flow size, flow duration, and flow rate for the Bell Labs I trace

	Minimum	1st Quartile	Median	Mean	3rd Quartile	Maximum	Standard Deviation
Size	1	283	604	27 980	4 641	3.941×10^9	4 768 760
Duration	2.724	79.95	266.8	37 920	2 105	3.193×10^8	955 486.4
Rate	1.865×10^{-7}	0.720	3.211	19.16	11.36	1 331	55.881
ln(*Size*)	0	5.645	6.404	7.084	8.443	22.090	1.875
ln(*Duration*)	1.002	4.381	5.587	6.255	7.652	19.580	2.467
ln(*Rate*)	−15.490	−0.328	1.167	0.829	2.431	7.194	2.495

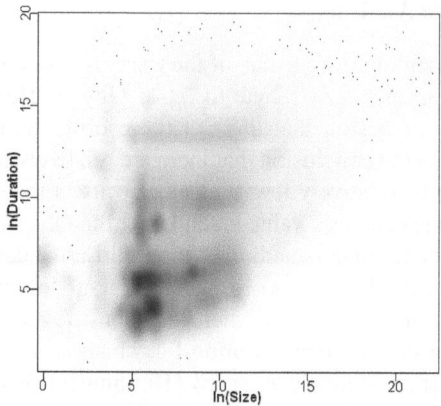

Fig. 3. Density plot of the logarithm of flow size (bytes) versus logarithm of flow duration (ms)

4.2 Means, Standard Deviations and Correlations

We have estimated the parameters of the truncated bivariate normal distribution of the logarithms of flow size, $\ln(Si)$, and flow duration, $\ln(Du)$. The values of the estimates obtained are given in Table 2. Comparing these values to the corresponding ones presented in Table 1 (obtained when we ignore the distributional assumption on the observed variables) we can conclude that they seem similar in the logarithmic scale, but turn out to be quite different in the original scale. Note that the values presented in Table 1 are empirical estimates of the characteristics of U, V, and W i.e. the logarithms of size, duration and rate in the truncated domain, while the values in Table 2 corresponds to estimates of the moments of these quantities in the non-truncated domain.

As we want to measure the dependence between flow size and flow duration (in the logarithmic scale), we start by ignoring the distributional assumption, and compute three of the most popular measures of association between the observed data (concretization of the truncated variables), presented in Table 3: Pearson's, Spearman's, and Kendall's tau correlation coefficients. For more information about measures of association see [14]. Both Spearman's and Kendall's tau correlation coefficients do not depend on the exact observed values, but only on their ranks, and are, therefore, invariant under the logarithmic transformation. However, the same is not true for Pearson's correlation coefficient, whose estimate between size and duration in the logarithmic scale is 0.365. Concurrently, we have used the truncated model to obtain the

Table 2. Estimates based on the truncated model for the mean and standard deviation of flow characteristics

X	$\hat{\mu} = \hat{E}(X)$	$e^{\hat{\mu}}$	$\hat{\sigma} = \sqrt{\hat{\mathrm{Var}}(X)}$
$\ln(Size)$	7.104	1216.394	1.828
$\ln(Duration)$	6.085	439.066	2.306
$\ln(Rate)$	1.019	2.770	2.533

Table 3. Estimates of the correlation coefficient

	Pearson				Truncated
	Original scale	Logarithmic scale	Kendall	Spearman	Model
Corr $(\widehat{Size, Duration}) = \hat{\rho}$	0.069	0.365	0.241	0.352	0.266
Corr $(\widehat{Size, Rate})$	0.012	0.391	0.302	0.414	0.479
Corr $(\widehat{Duration, Rate})$	−0.013	−0.714	−0.457	−0.631	-0.718

estimate of the correlation coefficient of the logarithms of flow size and flow duration in an hypothetical situation where no limitation on the duration and rate were imposed (non-truncated domain). The estimated value is presented in the last column of Table 3, which has the value 0.266. Thus, the estimated linear association between $\ln(Si)$ and $\ln(Du)$ (Pearson's correlation coefficient) in the domain where physical limitations to rate and duration are taken into account is larger than the value obtained when no limitations are considered. This is due to the form of the truncated domain and the fact that the correlation between $\ln(Si)$, and $\ln(Du)$ is positive. The reverse effect is observed for the pair $(\ln(Si), \ln(Ra))$. For $(\ln(Du), \ln(Ra))$ small differences are obtained between the estimates of the linear measure of association in the two domains.

4.3 Estimated Regression Equations

In order to estimate the expected behaviour of the logarithm of the size for flows with a given duration, $E(U|V = y)$, we use the result (8) and plug in the estimates of the parameters obtained as described in Section 3.2 (see tables 2 and 3). The obtained estimated regression equation is represented in Figure 4. In order to compare these results to the ones obtained ignoring truncation, we have added the estimated regression equation $\hat{E}(X|Y = y)$ considering $(X, Y) \sim N\left(\mu_1, \mu_2, \sigma_1^2, \sigma_2^2, \rho\right)$ and the parameters are estimated as in the previous case. Figure 4 shows also the estimated regression lines using least squares, LS, whose estimates of the slope, $\hat{\beta}_1$, and intercept, $\hat{\beta}_0$, are presented in Table 4. We use these alternative methods to understand what happens when truncation is not considered. A similar study was done to understand the expected behaviour of the logarithm of the duration for flows with a given size, $E(V|U = x)$. The results obtained are shown in Figure 5 and Table 5. From this figure it becomes clear that since our model takes into account that the rates are upper-bounded, the estimated regression equation, $\hat{E}(V|U = x)$, is the only one capable of estimating coherently the logarithm of duration for flows with large sizes.

Table 4. Estimates of the parameters, $\hat{\beta}_0$ and $\hat{\beta}_1$, of the estimated regression lines, $\hat{E}(\ln(Si)|\ln(Du) = y)$

Method	$\hat{\beta}_0$	$\hat{\beta}_1$
$\hat{E}(\ln(Si)|\ln(Du) = y)$	5.816	0.211
Least Squares	5.349	0.277

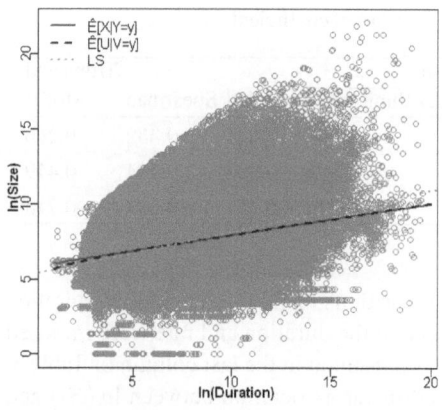
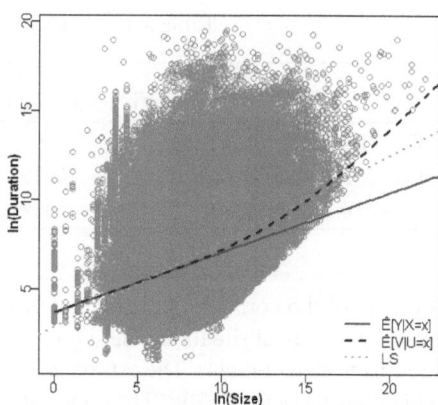

Fig. 4. Estimates of the expected value of the logarithm of flow size given the logarithm of flow duration and the LS regression line

Fig. 5. Estimates of the expected value of the logarithm of flow duration given the logarithm of flow size and the LS regression line

Table 5. Estimates of the parameters, $\hat{\beta}_0$ and $\hat{\beta}_1$, of the estimated regression lines, $\hat{E}\left(\ln\left(Du\right)|\ln\left(Si\right)=x\right)$

Method	$\hat{\beta}_0$	$\hat{\beta}_1$
$\hat{E}\left(\ln\left(Du\right)|\ln\left(Si\right)=x\right)$	3.704	0.336
Least Squares	2.853	0.480

Table 6. Estimates of the parameters, $\hat{\beta}_0$ and $\hat{\beta}_1$, of the estimated regression lines, $\hat{E}\left(\ln\left(Ra\right)|\ln\left(Du\right)=y\right)$

Method	$\hat{\beta}_0$	$\hat{\beta}_1$
$\hat{E}\left(\ln\left(Ra\right)|\ln\left(Du\right)=y\right)$	5.816	-0.789
Least Squares	5.349	-0.723

We have also studied the expected behaviour of the logarithm of the transmission rate for flows with a given size and flows with a given duration using LS along with (4) and (5). The results are shown in figures 6 and 7 and in tables 6 and 7. Figure 7 clearly illustrates that the estimates of the expected values of the rate for flows with a given size obtained ignoring truncations propose higher values than the maximum transmission rate admissible by the network.

As $p = P\{(X, Y) \in D\}$ is given by (7), we estimated this probability using the estimates of the parameters obtained as described in Section 3.2, getting $\hat{p} = 0.999996$. As this value is really close to 1, it is easy to justify why some of the estimates obtained for the parameters considering truncation and not considering truncation are close in the logarithmic scale but not so similar when transformed into the original domain.

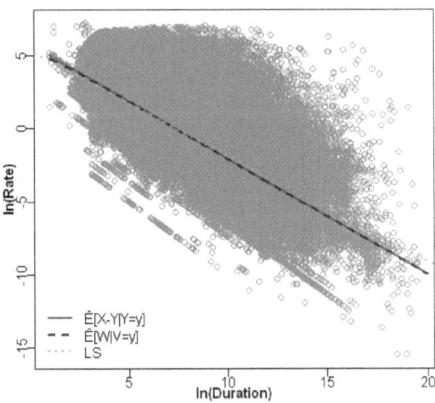 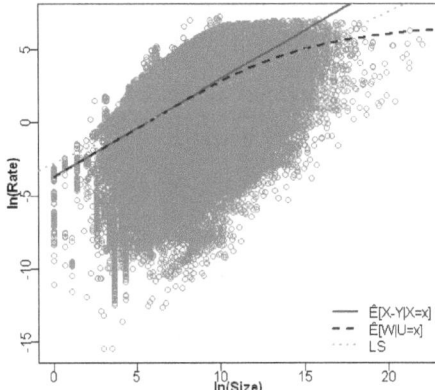

Fig. 6. Estimates of the expected value of the logarithm of flow rate given the logarithm of flow duration and the LS regression line

Fig. 7. Estimates of the expected value of the logarithm of flow rate given the logarithm of flow size and the LS regression line

Table 7. Estimates of the parameters, $\hat{\beta}_0$ and $\hat{\beta}_1$, of the estimated regression lines, $\hat{E}\left(\ln\left(Ra\right)|\ln\left(Si\right)=x\right)$

Method	$\hat{\beta}_0$	$\hat{\beta}_1$
$\hat{E}\left(\ln\left(Ra\right)|\ln\left(Si\right)=x\right)$	-3.704	0.664
Least Squares	-2.853	0.520

5   Conclusions

In this paper, we proposed a new model for studying the dependencies observed in Internet flow characteristics. We considered a bivariate lognormal distribution for the flow size and flow duration, and derived estimators for the mean, the variance, and the correlation based on a truncated domain that reflects the existence of upper-bounds on the duration and rate of measured flows. We have concluded that estimates of the correlation between the logarithm of the size and the logarithm of the duration are smaller in an hypothetical situation where no physical limitations (non-truncated domain) on size or rate are imposed. The form of the truncation domain, D, and a positive correlation between the logarithm of the size and the logarithm of the duration seems to be the two major reason for this fact. The strongest association observed is between the logarithm of the rate and the logarithm of the duration, where the estimates obtained considering or ignoring truncation are similar. This fact is also illustrated by similar values obtained for the various estimates of the expected value of the logarithm of the rate given a certain duration (as illustrated by Figure 6 and Table 6). As expected, the two variables are negatively correlated, i.e. long flows have low transmition rates (in the logarithmic scale).

Finally, we derived regression equations that describe the expected value of one characteristic (size, duration or rate) given other (size or duration). The estimates of these equations provide further insights on the dependencies between Internet flow characteristics. In particular, we have shown that, for flows with large sizes, our regression equations are able to predict durations and rates that are coherent with the physical limitations imposed by the network.

References

1. Hernandez-Campos, F., Marron, J.S., Resnick, S.I., Park, C., Jeffay, K.: Extremal dependence: Internet traffic applications. Stochastic Models 22(1), 1–35 (2005)
2. Kilpi, J., Lassila, P., Muscariello, L.: On dependence of Internet flow traffic. In: Proceedings of Second EuroNGI, Workshop on New Trends in Modelling, Quantitative Methods and Measurements, Aveiro, Portugal, November 24-25 (2006)
3. Lan, K.-C., Heidemann, J.: A measurement study of correlation of Internet flow characteristics. Computer Networks 50(1), 46–62 (2006)
4. Kilpi, J., Markovich, N.M.: Bivariate statistical analysis of tcp-flow sizes and durations. Euro-FGI Deliverable D.WP.JRA.5.1.1, New Mathematical Methods, Algorithms and Tools for Measurement, IP Traffic Characterization and Classification (2006)
5. Zhang, Y., Breslau, L., Paxson, V., Shenker, S.: On the characteristics and origins of Internet flow rates. In: Proceedings of ACM SIGCOMM Conference, vol. 32(4), pp. 309–322 (2002)
6. Barford, P., Crovella, M.: Generating Representative Web Workloads for Network and Server Performance Evaluation. In: Proceedings of SIGMETRICS 1998 (1998)
7. Ivanov, V.V., Znelov, P.V.: On a log-normal distribution of network traffic. Physica, 72–85 (2002)
8. Lakhina, A., Papagiannaki, K., Crovella, M., Diot, C., Kolaczyk, E.D., Taft, N.: Strutural analysis of network traffic flows. In: Proceedings of SIGMETRICS/Performance 2004, pp. 61–72 (2004)
9. Balakrishnan, H., Seshan, S., Stemm, M., Katz, R.: Analyzing stability in wide-area network performance. In: Proceedings of ACM SIGMETRICS 1997 (1997)
10. Pascoal, C.: Distribuição normal bivariada truncada com aplicação ao estudo de fluxos de tráfego de Internet, M.S. thesis, Instituto Superior Técnico, Technical University of Lisbon, Lisbon (2007)
11. Johnson, N.L., Kotz, S.: Continuous Univariate Distributions, vol. II. Wiley, New York (1995)
12. Khatri, C.G., Jaiswal, M.C.: Estimation of parameters of a truncated bivariate normal distribution. Journal of American Statistical Association 58(302), 519–526 (1963)
13. R Development Core Team, R: A Language and Environment for Statistical Computing, R Foundation for Statistical Computing, Vienna, Austria (2008), ISBN 3-900051-07-0
14. Croux, C., Filzmoser, P.: Projection pursuit based measures of association, TS-03-3. 200X

Peer-Level Analysis of Distributed Multimedia Content Sharing

Joana Gonçalves, Paulo Salvador, António Nogueira, and Rui Valadas

University of Aveiro/Instituto de Telecomunicações
3810-193 Aveiro, Portugal
{joanabg,salvador,nogueira,rv}@ua.pt

Abstract. In the last few years, peer-to-peer (P2P) file-sharing applications have become very popular: more users are continuously joining such systems and more objects are being made available, seducing even more users to join. Today, the traffic generated by P2P systems accounts for a major fraction of the Internet traffic and is bound to increase. P2P networks are currently evolving towards a real time multimedia content distribution able to provide reliable IPTV and Video on Demand (VoD) services. These new services create several challenges, the most important one being how to efficiently share resources in an heterogeneous, dynamic and continuously evolving group of peers. A detailed peer-level characterization is therefore fundamental for the technical and marketing design, deployment and management of these new services.

This paper will be focused on the characterization of BitTorrent as one of the most important P2P applications and, from that analysis, will try to extrapolate the main characteristics and behaviours of the potential P2P real-time multimedia content distribution clients. Specifically, the paper analyses the geographical localization of the involved peers, the peers' availability and the Round Trip Time (RTT) between peers as a function of the time and for different categories of shared files.

1 Introduction

Peer-to-Peer networks are computer connections through the Internet that allow users to share their resources, such as computation power, data and bandwidth, in such a way that they act both as clients and servers simultaneously. Nowadays, these networks are very popular and are responsible for a significant fraction of the Internet traffic, mainly because of their good architecture characteristics, like scalability, efficiency and performance. The P2P paradigm proved to be much more efficient than client-server communication especially for fast distribution of large amounts of data, since bottlenecks at servers with sporadic popularity are avoided by distributing requested data and the available access capacity over a global community of recipients.

Peer-to-peer (P2P) networks are currently one of the major platforms for non-real-time multimedia content distribution. However, P2P networks are evolving towards a real time multimedia content distribution able to provide reliable IPTV and Video on Demand (VoD) services. Recently, several Peer-to-peer (P2P) video streaming systems have been presented [1, 2, 3, 4] and studies on the potential of P2P systems as valuable

R. Valadas and P. Salvador (Eds.): FITraMEn 2008, LNCS 5464, pp. 81–95, 2009.
© Springer-Verlag Berlin Heidelberg 2009

and efficient platforms for IPTV and VoD services were also conducted [5]. These new services create several challenges, the most important one being how to efficiently share resources in an heterogeneous, dynamic and continuously evolving group of peers. A detailed peer-level characterization is therefore fundamental to the technical design, deployment and management of these new services. Moreover, this analysis will also be vital for the definition of the commercial and marketing plans of these new services.

An accurate characterization of these networks must rely on a good knowledge of the P2P applications and protocols, which can be a hard task due to the constant evolution of these protocols and the continuous appearance of new solutions. In this paper, we will focus our attention on the characterization of BitTorrent, currently one of the most popular P2P networks. We intend to analyze several peer-level characteristics of the BitTorrent community, for different types of multimedia shared files: the geographical localization of the involved peers, the temporal evolution of the involved peers' availability and the temporal variability of the Round Trip Time (RTT) between peers. From those results, we expect to extrapolate the main characteristics and behaviours of the potential P2P real-time multimedia content distribution clients.

The paper is organized as follows: section 2 presents some related work on this subject; section 3 briefly describes the methodology that was used for this study; section 4 presents and discusses the results obtained for each one of the studied topics: geographical distribution of peers, peers' availability and round trip time between involved peers; finally, section 5 presents the main conclusions and outlines some topics for future research.

2   Related Work

Due to the significant growth of P2P networks over the last few years, some works [6, 7, 8, 9, 10, 11, 12] have been done in order to better understand how they work and, thus, allowing to optimize their performance and avoid future overload problems on Internet traffic. Being BitTorrent [13] one of the most popular P2P applications, it is also a big target of evaluation. Therefore, studies on modelling and analysis of BitTorrent applications have been done aiming to improve the performance and to characterize these systems at a peer level. In this way, some mathematical models for BitTorrent were proposed in [14, 15, 16, 17, 18]. In [14] a simple fluid model for the BitTorrent application is presented and the steady state network performance was studied. In [15] a new strategy of peer's selection is proposed in order to turn download faster. With the aim of studying the performance of piece scheduling decisions made at the seed, a stochastic approximation was done in [16]. Reference [17] presents BitProbes, a system performing measurements to Internet end hosts, analysing their geographical distribution and upload capacity. Some changes in BitTorrent systems are suggested in [18] to facilitate efficient streaming between peers as well as providing soft Quality of Service guarantee. In [19,20,21,22,23] some simulation analysis were made, aiming to characterize BitTorrent traffic. In [22] the high level characteristics and the user behaviour of BitTorrent were tested by analysing the activity, availability, integrity, flash-crowd and download performance of this application. In [23], using the Multiprobe framework, measurements of BitTorrent and Internet were conducted and statistically correlated with location, route, connectivity and traffic. In [19], the peers' performance corresponding to the download of a unique torrent file of 1.77GB of content, the Linux Redhat 9 distribution, during 5 months was evaluated in terms of throughput per client during the download, the ability to sustain high flash-crowd and the geographical distribution of the involved peers. A geographical, organizational, temporal, network robustness and peer activity analysis of the BitTorrent application, for 120 files, was presented in [20]. In [21] several studies about the popularity, availability, content lifetime and injection time, download speed and pollution level for this application were presented. In [24] a similar study was conducted focusing on the geographical distribution and availability of BitTorrent, but it was restricted only to video files and comprised a shorter period of time - 8 days.

3   Methodology

The measurements that constitute the basis for this study were obtained from January to August 2008 in two distinct Portuguese cities, Aveiro and Coimbra, and using different Internet connections on each case: a CATV 12 Mbps Internet access and an ADSL 4 Mbps Internet access, respectively. Vuze (formerly known as Azureus) [25] was used to download six different categories of files: 2008 Movies, Music, Animated movies, French movies, Asian movies and Linux distribution (used as a comparison to multimedia files), involving a total of 25 files and 85293 peers. The selection of the files was made according to the file ranking for each category, in terms of the number of involved peers, by choosing the files that were positioned on the top of the list.

For each torrent, the identification of the peers' IP addresses and application TCP ports was achieved by analysing the BitTorrent distributed hash table (DHT) maintenance messages, namely, the announce messages with peer's information. The next step was the localization of the country corresponding to the IP address of each peer, using the Geoiplookup tool [26].

In order to obtain the Round Trip Time between hosts, the IP address and TCP port number of the involved peers was used to execute periodic TCP port scans with Nmap [27]. From these scan tests, it was possible to determine the peer's availability, which can be defined as the percentage of peers to whom it was possible to establish a TCP connection.

4   Results

4.1   Geographical Distribution

This section will analyse and evaluate the geographical distribution of peers around the whole world. It will start with an analysis of the distribution of peers per continent and per country, mainly those countries having the higher number of identified peers. Then, in order to obtain a more significant and meaningful result about the importance of this P2P system, an analysis of the distribution of identified peers normalized by their country population is done.

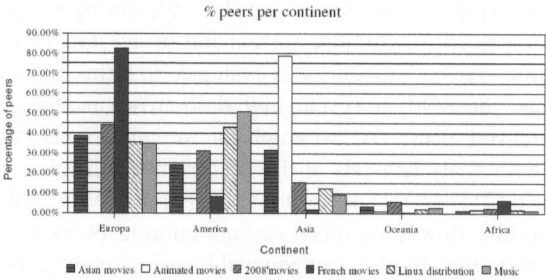

Fig. 1. Percentage of peers per continent of all categories

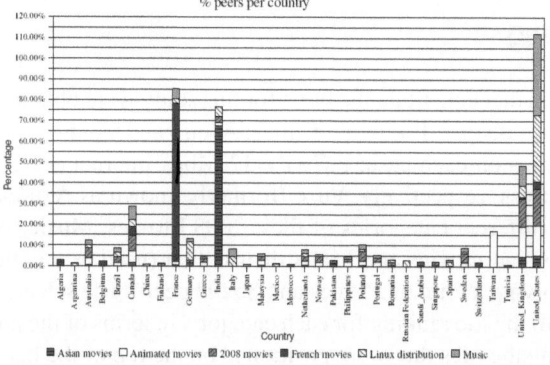

Fig. 2. Percentage of peers per country of all categories

Figures 1 and 2 present the results obtained for the six categories that correspond to the different types of shared files. Countries/Continents with a higher degree of technological development and larger population have a high number of peers. From Figure 2, one can clearly see the high percentage of peers that United States of America involves, which can be explained by several factors: the cultural diversity of USA, making people to have different characteristics and interests; the technological development of the country, with good networks infrastructures and easy Internet access for the major portion of the population; the economical level of the country, turning Internet access into a cheap service and contributing to the wide spread of the technology advances. It is important to mention that categories where USA is not in the first position are those that correspond to files specific to other countries: French and Asian movies categories, for example, where obviously France and India, respectively, have the domain on the number of involved peers. The small use of BitTorrent on China can be explained by the fact that the considered files did not raise a big interest from Chinese users, which usually prefer Chinese dubbed movies whose torrents are indexed in websites that are completely distinct form the ones considered in these study.

In order to obtain a more significant and meaningful result about the importance of this P2P system on each country, an analysis of the distribution of identified peers normalized by their country population is done. Figures 3 to 8 show normalized maps for

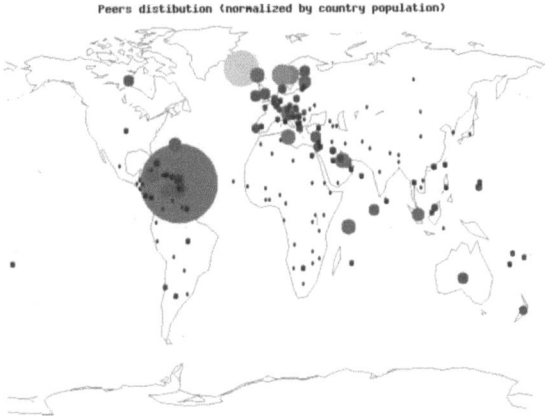

Fig. 3. Peers' distribution normalized by the country population - 2008 movies category

Fig. 4. Peers' distribution normalized by the country population - music category

each studied category. It is obviously expected that the larger is the country population, the higher will be the probability of having more peers involved. The fall of USA after this normalization is clearly visible, allowing us to conclude that the high number of peers that this country involves is not so relevant when compared to its population, which reaches almost 5% of the world population. On the other hand, it is necessary to refer that some countries have such a low number of inhabitants that, after normalization is done, they take a place on the top of the ranking although that fact does not directly mean that they are extremely relevant countries on this way of sharing files. By the analysis that was made, we can conclude that the localization of the more relevant peers is strongly correlated with the kind of file in analysis: for example, for the French movies category countries with more interest were those with some relationship with the French language; for the Asian movies category, peers are more concentrated in the Asian continent. The relative importance of Canada, United Kingdom and

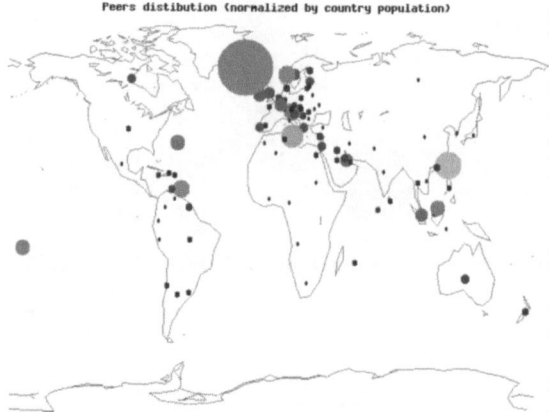

Fig. 5. Peers' distribution normalized by the country population - animated movies category

Fig. 6. Peers' distribution normalized by the country population - French movies category

Australia in this P2P file sharing system is also quite obvious, which is not surprising since all of these countries are economically developed, with good network infrastructures and easy Internet access. When analysing the normalized number of peers it is possible to observe that for the 2008 Movies, Animated Movies and Music categories, Nordic and Central Europe countries have an higher predominance of peers, which reveals a cultural aptness of these populations to consume multimedia contents at home, maybe due to climate constrains as suggested in [24]. Furthermore, it is important to remind that, except for the 2008 Movies and Asian movies categories, other categories have a small sample of peers involved, between 2000 and 5000, and in most of the cases they are strongly concentrated on only one country. This fact makes the desired analysis a little bit more difficult. For these categories the results obtained before and after normalization were similar, as opposed to the other categories that have much more peers involved.

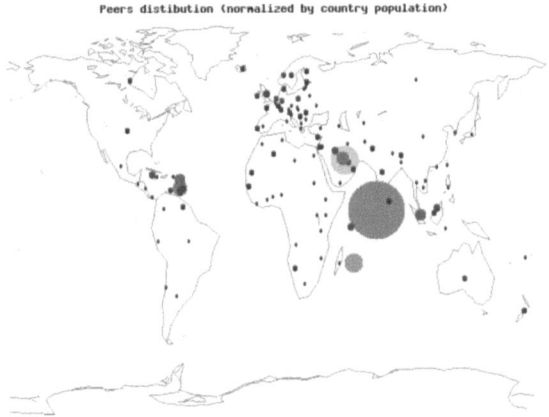

Fig. 7. Peers' distribution normalized by the country population - Asian movies category

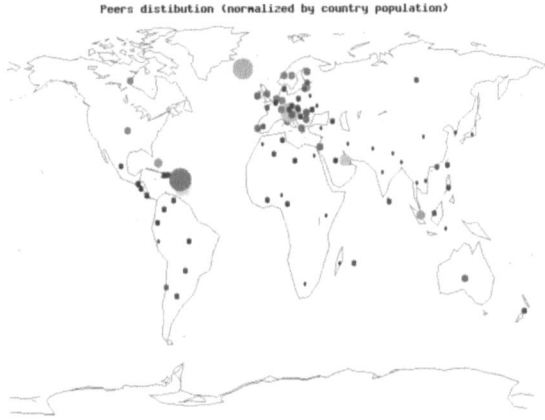

Fig. 8. Peers' distribution normalized by the country population - Linux distribution category

4.2 Availability

Figures 9, 10 and 11 depict the obtained results for the peers' availability per category, for different daily periods and for different days.

As can be observed, the peers' availability for almost all categories decreased very slowly during the period of study. This slow decline shows that, in general, people keep sharing files; it is not just a specific file that they are searching for. Therefore, we can conclude that P2P file sharing is becoming more and more common on people lives and not a sporadic attitude. From the above figures, we can also observe that the peers' availability percentage is never around 100%, starting in almost all cases with 25% of the whole peers sample except for the 2008 movies category that starts with 45% of the total number of peers. This can be justified by the fact that an increasing number of hosts are firewalled or located behind NAT boxes and proxies in order to

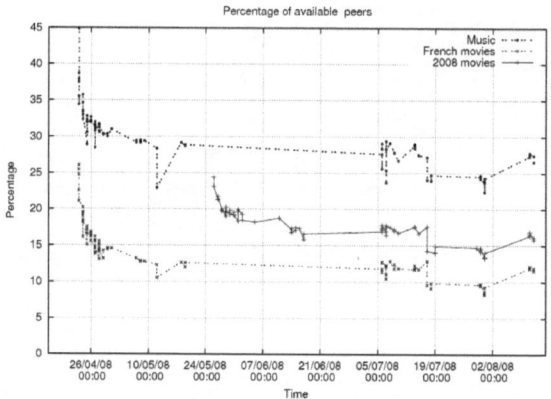

Fig. 9. Evolution of the peers' availability during the period of analysis - music, French movies and 2008 movies categories

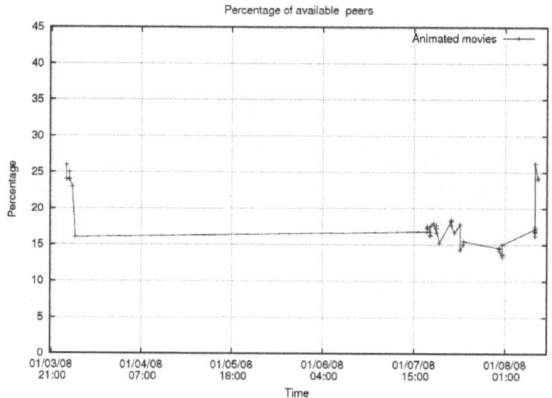

Fig. 10. Evolution of the peers' availability during the period of analysis - animated movies category

avoid detection, and consequently they do not answer to any download request or any probing test. Such conclusions can be taken since the probing tests were done just after the knowledge of the peers' ports: obviously, at least at that time instant a great portion of those peers should be available and, if not, that is because such ports are protected against intrusions.

With the aim of evaluating the relationship between the peers' availability and the time of the day, an analysis of such availability was done by distributing peers by their respective countries, allowing us to pay attention to the time zone of each case. Countries with similar time zone were grouped on the same plot. Figures 12 to 16 present the evaluation of the peers' availability for the most important involved countries in terms of the number of peers. Note that the time zone represented on these plots is the GMT time.

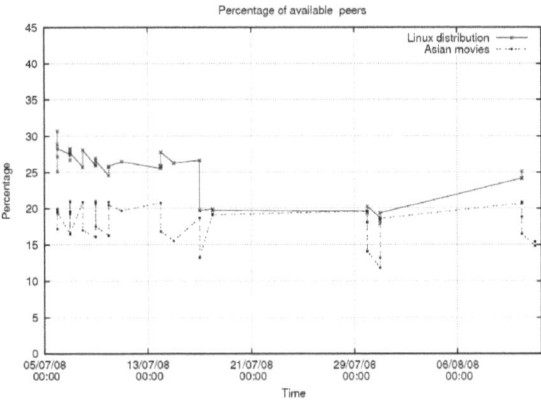

Fig. 11. Evolution of the peers' availability during the period of analysis - Asian movies category

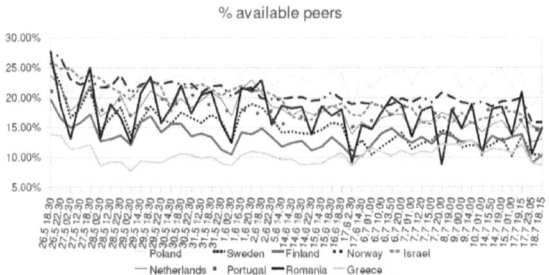

Fig. 12. Percentage of available peers during the period of analysis for peers from Israel and eight European countries

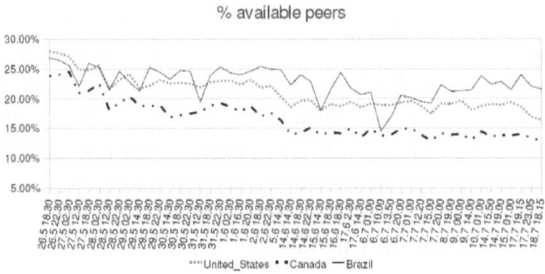

Fig. 13. Percentage of available peers during the period of analysis for peers from United States, Canada and Brazil

By observing peers' availability during the daily period, we can see that there are not huge changes, which can be explained by the fact that our analysis involved peers from all over the world, having different time zones according to their localization. Nonetheless, for the Indian movies category differences on the number of available

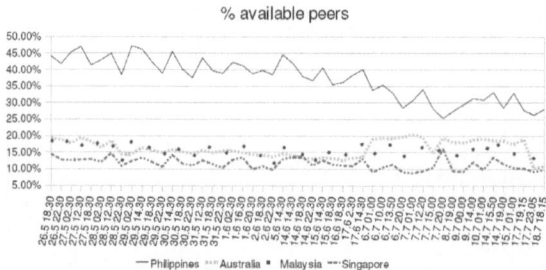

Fig. 14. Percentage of available peers during the period of analysis for peers localized on Philippines, Australia, Malaysia, Singapore

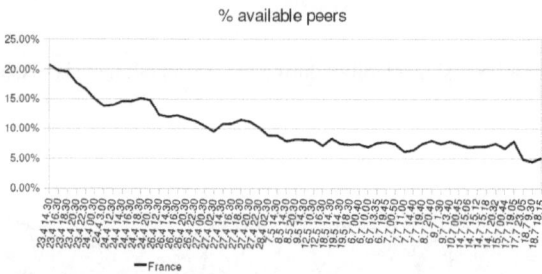

Fig. 15. Percentage of available peers during the period of analysis for peers localized in France

peers for different daily hours are visible. The daily period with the lowest values is, as we can observe, between 8.30 pm and 6.30 am, corresponding to the GMT time between 4.00 pm and 2.00 am. We can conclude that the period of time with lower number of users coincide with day transitions, which can be somehow justified: (i) people that does not have unlimited traffic nor any period of the day with free Internet access are encouraged to turn off their connections and computers when they are not needing it, for example when they are sleeping; (ii) a great portion of the companies are closed during the night and their employees' computers are turned off; (iii) Internet access in India is not cheap, being not accessible at home for a major slice of the population. On the other hand, it is also possible to observe that the higher values appear generally at afternoon, which is also comprehensive if we keep in mind that this is the daily period where the major portion of the population finishes their work, arrive home, turn on their computers and can finally start sharing files.

4.3   Round Trip Time

This section presents a study of the RTT variability in order to find out its dependences, especially with the distance between peers, Internet connection quality and time of the day. As was already mentioned, RTT is the time it takes for a packet to go from one peer to another and come back. Thus, variations are expected according to the distance that separates the origin from the end peer. In this way, firstly a study of such relationship between RTT and host distances was made. Polar maps showed on Figures 17

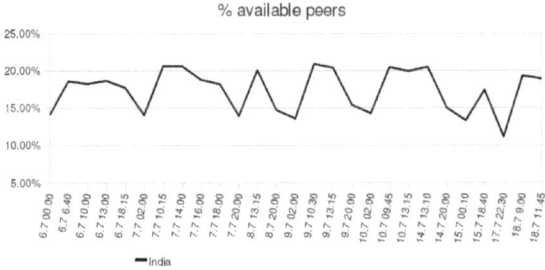

Fig. 16. Percentage of available peers during the period of analysis for peers localized in India

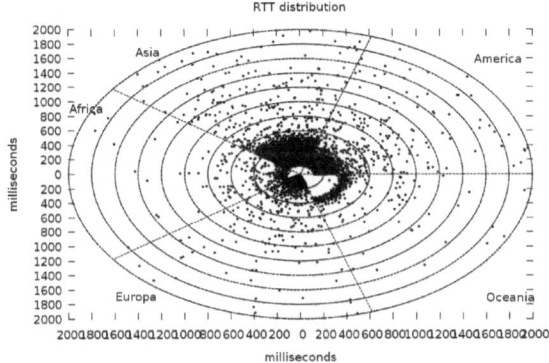

Fig. 17. Round Trip Time distribution for the CATV 12Mbps Internet connection

and 18 present the RTT distribution per continent for the 12 Mbps CATV and 4 Mbps ADSL Internet connections, respectively. As can be observed, European peers present the lower RTT values, which are concentrated at the center of the plot, from zero to the 200 ms circle, as was expected since the origin host is localized in Portugal. American and Asian peers have mostly of their RTT values located between 100 and 400 ms. Finally, Oceania RTTs are placed around 400 ms and even more. Unfortunately, due to the low number of African peers, it is not so easy to take any relevant conclusions about the RTT values for end hosts localized on this continent, but nevertheless it is possible to observe that they have similar RTT values to the ones corresponding to peers located in America and Asia, which is acceptable since the related distances are similar. From this analysis it is possible to conclude about the strong dependence of RTT from the distance between hosts. Besides, by comparing the polar maps we can easily see a much higher concentration of peers around 1000 and 1400 ms for the ADSL Internet connection, especially for peers localized in Europe but also in America and Asia. Oceania has also more peers around 1400 ms for the ADSL Internet connection, when compared to the CATV 12 Mbps Internet connection.

Since different connection types have different routes, this can explain such differences on RTT values. In order to obtain the different routes, a traceroute tool was used.

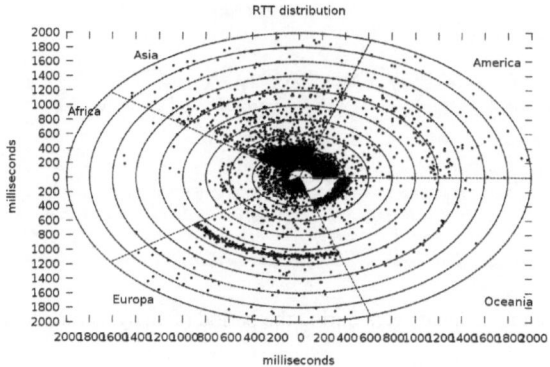

Fig. 18. Round Trip Time distribution for the ADSL 4Mbps Internet connection

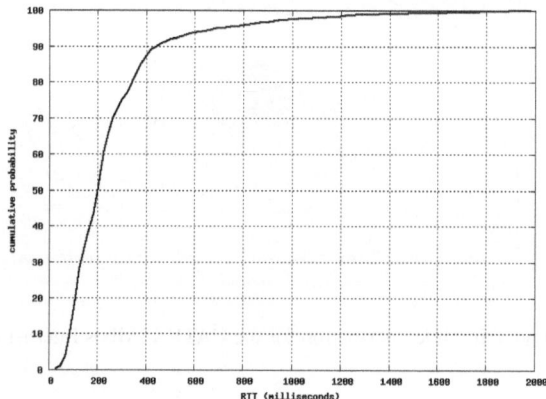

Fig. 19. RTT cumulative distribution function for the CATV 12Mbps Internet connection

When a route can not be traced and the corresponding RTT can not be measured, this tool also informs us on where and why the problem occurred, if it is a network problem or a router that is not working. From the analysis of the different routes for each Internet connection type, it was confirmed that in general CATV Internet connection provides more direct routes between origin and end hosts than ADSL Internet connection. In the ADSL Internet connection, for example, in order for a packet to reach any other European country it was observed that in most of the cases it passes through United Kingdom and presents longer routes when compared to the CATV Internet connection. This behavior can be explained by the fact that the ADSL provider has its headquarters located at UK and therefor all their international traffic transport is done through UK, in contrast with the CATV ISP that is a Portuguese-based provider. These facts explain the worst RTT results obtained for the ADSL Internet connection. Tests made for countries like France, Germany, Italy, Netherlands and United Kingdom proved the existence of this direct connection between hosts. For other continents, except Europe, the obtained traces were more mixed, since the best paths were not always given by the

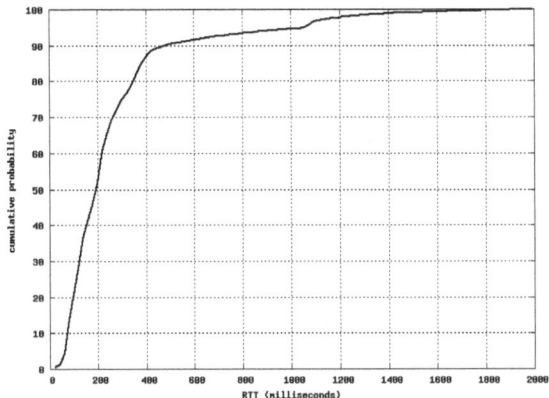

Fig. 20. RTT cumulative distribution function for the ADSL 4Mbps Internet connection

CATV Internet connection, although the passage through United Kingdom still keeps appearing in the ADSL Internet connections. Countries like Australia, Japan, Malaya, Morocco, Philippines, Taiwan, Turkey and United States have shorter routes when using the ADSL Internet connection.

Figures 19 and 20 present the RTT cumulative distribution functions for each connection type, enabling us to visualize the RTT probability evolution. It is possible to verify that RTT values are, in general, concentrated from 0 up to 400ms. Moreover, it is possible to observe that 10% of the peers have a RTT value of 80 msec or less and 20% of the peers have a RTT value of 100 msec or less. It is also possible to observe an higher concentration of RTT values on the 1050 to 2000 ms range for the ADSL Internet connection, when compared to the CATV Internet connection.

5   Conclusions

This paper focused on the peer-level characterization of specific and defining users' behaviors on a peer-to-peer multimedia file sharing environment. Specifically, the geographical localization of the involved peers, the temporal evolution of the involved peers' availability and the temporal variability of the Round Trip Time (RTT) between peers were studied and analyzed in detail. For this purpose, during an 8 months period, from January to August 2008, a BitTorrent application (Vuze) was used to identify peers sharing six different categories of files, involving a total of 25 files and 85293 peers.

From the analysis of the geographical distribution of the involved peers, some conclusions can be drawn: (i) P2P file sharing systems comprise a large number of countries, and the continents with more involved peers are Europe, America and Asia; (ii) United States of America, due to its population size, cultural diversity and social-economic development level is the country with more peers involved, in absolute numbers, but when normalized by their percentage of the world population, their position on the overall ranking decreases on some places; (iii) culture aptness constrained by climate characteristics plays a predominant role on the distribution of peers; (iv) peers distribution depends on the content category under analysis, proving that there are different interests for different countries.

After the geographical distribution, peers' availability was evaluated and some remarks can be made: (i) a strong number of peers is protected by firewall or are located behind NAT boxes and proxies, avoiding their detection; (ii) during all the analysis period, the peers' availability remains almost constant revealing a high level of user's loyalty to the sharing network.

Finally, concerning the obtained Round Trip Time results, it can be concluded that: (i) Round Trip Time is strongly dependent on the distance between the origin and destination hosts; (ii) different types of Internet connections lead to different route paths, resulting in visible differences on RTT results; (iii) distribution plots showed that the majority of the RTT values are located between 0 and 300 milliseconds, although 10% of the peers have a RTT value of 80 msec or less and 20% of the peers have a RTT value of 100 msec or less.

We strongly believe that the results obtained on this study, that should be necessarily complemented by more exhaustive studies, can be extrapolated to help on the deployment and management of new real-time multimedia content distribution services based on P2P architectures.

References

1. Akyol, E., Tekalp, A.M., Civanlar, M.R.: A flexible multiple description coding framework for adaptive peer-to-peer video streaming. IEEE Journal of Selected Topics in Signal Processing 1(2), 231–245 (2007)
2. Kim, D., Lee, J., Lee, Y., Kim, H.: AVS: An adaptive P2P video streaming scheme with frame type aware scheduling. In: The 9th International Conference on Advanced Communication Technology, February 12-14, vol. 2, pp. 1447–1452 (2007)
3. Lin, C.S., Cheng, Y.C.: 2MCMD: A scalable approach to VoD service over peer-to-peer networks. Journal of Parallel and Distributed Computing 67(8), 903–921 (2007)
4. Liu, Z., Shen, Y., Panwar, S.S., Ross, K.W., Wang, Y.: Using layered video to provide incentives in P2P live streaming. In: P2P-TV 2007: Proceedings of the 2007 workshop on Peer-to-peer streaming and IP-TV, pp. 311–316. ACM, New York (2007)
5. Sentinelli, A., Marfia, G., Gerla, M., Kleinrock, L., Tewari, S.: Will IPTV ride the peer-to-peer stream (peer-to-peer multimedia streaming). IEEE Communications Magazine 45(6), 86–92 (2007)
6. Lua, E.K., Crowcroft, J., Pias, M., Sharma, R., Lim, S.: A survey and comparison of peer-to-peer overlay network schemes. IEEE Communications Surveys and Tutorials 7(2), 72–93 (2005)
7. Karagiannis, T., Broido, A., Faloutsos, M., Claffy, K.: Transport layer identification of P2P traffic. In: Proc. ACM Sigcomm Internet Measurement Conference (October 2004)
8. Sen, S., Spatscheck, O., Wang, D.: Accurate, scalable in-network identification of P2P traffic using application signatures. In: Proc. of the 13th international Conference on World Wide Web, New York, NY, USA, May 17-20, pp. 512–521 (2004)
9. Saroiu, S., Gummadi, P.K., Gribble, S.D.: A measurement study of peer-to-peer file sharing systems. In: Proc. of Multimedia Computing and Networking (2002)
10. Leibowitz, N., Bergman, A., Ben-Shaul, R., Shavit, A.: A measurement study of peer-to-peer file sharing systems. In: Proc. of the 7th Int. WWW Caching Workshop (2002)

11. Bhagwan, R., Savage, S., Voelker, G.: Understanding availability. In: Kaashoek, M.F., Stoica, I. (eds.) IPTPS 2003. LNCS, vol. 2735, pp. 256–267. Springer, Heidelberg (2003)
12. Gkantsidis, C., Mihail, M., Saberi, A.: Random walks in peer-to-peer networks. In: Proc. of INFOCOM 2004, Twenty-third Annual Joint Conference of the IEEE Computer and Communications Societies (2004)
13. BitTorrent, http://www.bittorrent.com
14. Qiu, D., Srikant, R.: Modeling and performance analysis of bittorrent-like peer-to-peer networks. In: Proc. of ACM SIGCOMM 2004, Portland, Oregon, USA, August 30 - September 03, pp. 367–378 (2004)
15. Tian, Y., Wu, D., Ng, K.W.: Modeling, analysis and improvement for bittorrent-like file sharing networks. In: Proc. of INFOCOM 2006, 25th IEEE International Conference on Computer Communications (April 2006)
16. Michiardi, P., Ramachandran, K., Sikdar, B.: Modeling, analysis and improvement for bittorrent-like file sharing networks. In: Proc. of NETWORKING 2007, Ad Hoc and Sensor Networks, Wireless Networks, Next Generation Internet, 6th International IFIP-TC6 Networking Conference, Atlanta, GA, USA, May 14-18 (2007)
17. Isdal, T.: Msc. Thesis: Extending BitTorrent for Streaming Applications. The Royal Institute of Technology (KTH), Stockholm, Sweden (October 2006)
18. Erman, D.: Extending bittorrent for streaming applications. In: Proc. 4th Euro-FGI Workshop on New Trends in Modelling, Quantitative Methods and Measurements, Ghent, Belgium (2007)
19. Izal, M., Urvoy-Keller, G., Biersack, E., Felber, P., Hamra, A.A. GarcÃ©s-Erice, L.: Dissecting bittorrent: Five months in a torrent's lifetime. In: Proc. Passive and Active Network Measurement, April 2004, pp. 1–11 (2004)
20. Iosup, A., Garbacki, P., Pouwelse, J., Epema, D.: Analyzing bittorrent: Three lessons from one peer-level view. In: Proc. 11th ASCI Conference (2005)
21. Pouwelse, J., Garbacki, P., Epema, D., Sips, H.: An introduction to the bittorrent peer-to-peer file sharing system. In: Hand-out at the 19th IEEE Annual Computer Communications Workshop (October 2004)
22. Pouwelse, J., Garbacki, P., Epema, D., Sips, H.: The bittorrent P2P file-sharing system: Measurements and analysis. In: Castro, M., van Renesse, R. (eds.) IPTPS 2005. LNCS, vol. 3640, pp. 205–216. Springer, Heidelberg (2005)
23. Iosup, A., Garbacki, P., Pouwelse, J., Epema, D.: Correlating topology and path characteristics of overlay networks and the internet. In: Sixth IEEE International Symposium on Cluster Computing and the Grid Workshops, May 16-19 (2006)
24. Salvador, P., Nogueira, A.: Study on geographical distribution and availability of bittorrent peers sharing video files. In: 12th Annual IEEE International Symposium on Consumer Electronics (ISCE 2008), Vilamoura, Algarve, Portugal, April 14-16 (2008)
25. Vuze, http://www.vuze.com
26. MaxMind - GeoLite Country - Open Source IP Address to Country Database, http://www.maxmind.com/app/geolitecountry
27. NMAP free security scanner for network exploration and hacking, http://nmap.org

Volume Anomaly Detection in Data Networks: An Optimal Detection Algorithm vs. the PCA Approach

Pedro Casas[1,3], Lionel Fillatre[2], Sandrine Vaton[1], and Igor Nikiforov[2]

[1] TELECOM Bretagne, Computer Science Department,
Technopôle Brest-Iroise, 29238 Brest, France
{pedro.casas,sandrine.vaton}@telecom-bretagne.eu
[2] Charles Delaunay Institute/LM2S, FRE CNRS 2848,
Université de Technologie de Troyes,
12 rue Marie Curie, 10010 Troyes, France
{lionel.fillatre,igor.nikiforov}@utt.fr
[3] Universidad de la República, Faculty of Engineering,
Julio Herrera y Reissig 565, 11300 Montevideo, Uruguay

Abstract. The crucial future role of Internet in society makes of network monitoring a critical issue for network operators in future network scenarios. The Future Internet will have to cope with new and different anomalies, motivating the development of accurate detection algorithms. This paper presents a novel approach to detect unexpected and large traffic variations in data networks. We introduce an optimal volume anomaly detection algorithm in which the anomaly-free traffic is treated as a nuisance parameter. The algorithm relies on an original parsimonious model for traffic demands which allows detecting anomalies from link traffic measurements, reducing the overhead of data collection. The performance of the method is compared to that obtained with the Principal Components Analysis (PCA) approach. We choose this method as benchmark given its relevance in the anomaly detection literature. Our proposal is validated using data from an operational network, showing how the method outperforms the PCA approach.

Keywords: Network Monitoring and Traffic Analysis, Network Traffic Modeling, Optimal Volume Anomaly Detection.

1 Introduction

Bandwidth availability in nowadays backbone networks is large compared with today's traffic demands. The core backbone network is largely over-provisioned, with typical values of bandwidth utilization lower than 30%. The limiting factor in terms of bandwidth is not the backbone network but definitely the access network, whatever the access technology considered (ADSL, GPRS/EDGE/UMTS, WIFI, WIMAX, etc.). However, the evolution of future access technologies and the development of optical access networks (Fiber To The Home technology) will

R. Valadas and P. Salvador (Eds.): FITraMEn 2008, LNCS 5464, pp. 96–113, 2009.
© Springer-Verlag Berlin Heidelberg 2009

dramatically increase the bandwidth for each end-user, stimulating the proliferation of new "bandwidth aggressive" services (High Definition Video on Demand, interactive gaming, meeting virtualization, etc.). Some authors forecast a value of bandwidth demand per user as high as 50 Gb/sec in 2030. In this future scenario, the assumption of "infinite" bandwidth at the backbone network will no longer be applicable. ISPs will need efficient methods to engineer traffic demands at the backbone network. This will notably include a constant monitoring of the traffic demand in order to react as soon as possible to abrupt changes in the traffic pattern. In other words, network and traffic anomalies in the core network will represent a crucial and challenging problem in the near future.

Traffic anomalies are unexpected events in traffic flows that deviate from what is considered as normal. Traffic flows within a network are typically described by a traffic matrix (TM) that captures the amount of traffic transmitted between every pair of ingress and egress nodes of the network, also called the Origin Destination (OD) traffic flows. These traffic flows present two different properties or behaviors: on one hand, a stable and predictable behavior due to usual traffic usage patterns (e.g. daily demand fluctuation); on the other hand, an abrupt and unpredictable behavior due to unexpected events, such as network equipment failures, flash crowd occurrences, security threats (e.g. denial of service attacks), external routing changes (e.g. inter-AS routing through BGP) and new spontaneous overlay services (e.g. P2P applications). We use the term "volume anomaly" [18] to describe these unexpected network events (large and sudden link load changes). Figure 1 depicts the daily usual traffic pattern together with sporadic volume anomalies in four monitored links from a private international Tier-2 network. As each OD flow typically spans multiple network links, a volume anomaly in an OD flow is simultaneously visible on several links. This multiple evidence can be exploited to improve the detection of the anomalous OD flows. Volume anomalies have an important impact on network performance, causing sudden situations of strong congestion that reduce the network throughput and increase network delay. Even more, in the case of volume network attacks, the cost associated with damages and side effects can be excessively high to the network operator. The early and accurate detection of these anomalies allows to rapidly take precise countermeasures, such as routing reconfiguration in order to mitigate the impact of traffic demands variation, or more precise anomaly diagnosis by deeper inspection of other types of traffic statistics.

There are at least two major problems regarding current anomaly detection in OD traffic flows: (i) most detection methods rely on highly tuned data-driven traffic models that are not stable in time [18,22] and so are not appropriate for the task, causing lots of false alarms and missing real anomalies; (ii) current detection methods present a lack of theoretical support for their optimality properties (in terms of detection rate, false alarm generation, delay of detection, etc.), making it almost impossible to compare their performances. In this context, there are many papers with lots of new anomaly detection algorithms that claim to have the best performance so far, but the generalization of these results is not plausible without the appropriate theoretical support. In this paper we focus on

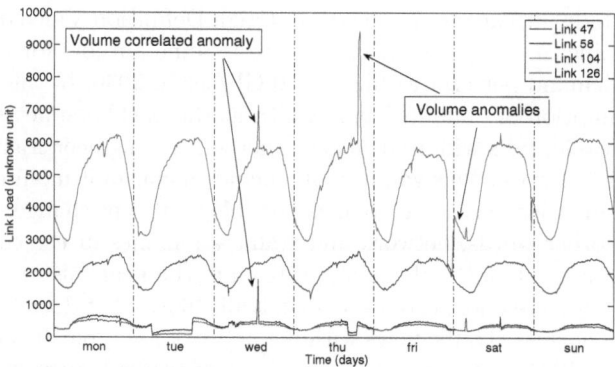

Fig. 1. Network anomalies in a large Tier-2 backbone network

the optimal detection of volume traffic anomalies in the TM. We present a new linear and parsimonious model to describe the TM. This model remains stable in time, making it possible to overcome the stability problems of different current approaches. At the same time, it allows to monitor traffic flows from simple link load measurements, reducing the overhead of direct flow measurements. Based on this model, we introduce a simple yet effective anomaly detection algorithm. The main advantages of this algorithm rest on its optimality properties in terms of detection rate and false alarm generation.

1.1 Related Work

The problem of anomaly detection in data networks has been extensively studied. Anomaly detection consists of identifying patterns that deviate from the normal traffic behavior, so it is closely related to traffic modeling. This section overviews just those works that have motivated the traffic model and the detection algorithm proposed in this work. The anomaly detection literature treats the detection of different kinds of anomalous behaviors: network failures [8,9,10], flash crowd events [11,12] and network attacks [15,16,17,19,25]. The detection is usually performed by analyzing either single [13,14,15,23] or multiple time-series [19,18,26], considering different levels of data aggregation: IP flow level data (IP address, packet size, nº of packets, inter-packet time), router level data (from router's management information), link traffic data (SNMP measurements from now on) and OD flow data (i.e a traffic matrix). The usual behavior of traffic data is modeled by several approaches: spectral analysis, Principal Components Analysis (PCA), wavelets decomposition, autoregressive integrated moving average models (ARIMA), etc. [13] analyzes frequency characteristics of network traffic at the IP flow level, using wavelets filtering techniques. [19] analyses the distribution of IP flow data (IP addresses and ports) to detect and classify network attacks. [15] uses spectral analysis techniques over TCP traffic for denial of service detection. [14] detects anomalies from SNMP measurements, applying exponential smoothing and Holt-Winters forecasting techniques. In [18], the authors use the Principal Components Analysis technique to separate the SNMP measurements in anomalous and anomaly-free traffic. These methods can detect anomalies by monitoring links traffic but they do not appropriately exploit the spatial correlation induced by the routing process. This correlation represents a key feature that can be used to provide more robust results. Moreover, the great majority of them cannot be applied when the routing matrix varies in time (because links traffic distribution changes without a necessary modification in OD flows), and many of the developed techniques are so data-driven that they are not applicable in a general scenario (notably the PCA approach in [18], as mentioned in [22]).

The authors in [26] analyze traffic at a higher aggregation level (SNMP measurements and OD flow data), using ARIMA modeling, Fourier transforms, wavelets and PCA to model traffic evolution. They extend the anomaly detection field to handle routing changes, an important advantage with respect to previous works. Unfortunately, all these methods present a lack of theoretical results on their optimality properties, limiting the generalization of the obtained results. [24] considers the temporal evolution of the TM as the evolution of the state of a dynamic system, using prediction techniques to detect anomalies, based on the variance of the prediction error. The authors use the Kalman filter technique to achieve this goal, using SNMP measurements as the observation process and a linear state space model to capture the evolution of OD flows in time. Even though the approach is quite appealing, it presents a major drawback: it depends on long-time periods of direct OD flow measurements for calibration purposes, an assumption which can be too restrictive in a real application, or directly infeasible for networks without OD flow measurement technology.

Our work deals with volume anomalies, i.e. large and sudden changes in OD flows traffic, independently of their nature. As direct OD flow measurements are rarely available, the proposed algorithm detects anomalies in the TM from links traffic data and routing information. This represents quite a challenging task: as the number of links is generally much smaller than the number of OD flows, the TM process is not directly observable from SNMP measurements. To solve this observability problem, a novel linear parsimonious model for anomaly-free OD flows is developed. This model makes it possible to treat the anomaly-free traffic as a nuisance parameter, to remove it from the detection problem and to detect the anomalies in the residuals.

1.2   Contributions of the Paper

This paper proposes an optimal anomaly detection algorithm to deal with abrupt and large changes in the traffic matrix. In [1] we present an optimal "sequential" algorithm to treat this problem, minimizing the anomaly detection delay (i.e. the time elapsed between the occurrence of the anomaly and the rise of an alarm). In this work, we draw the attention towards a "non-sequential" detection algorithm. This algorithm is optimal in the sense that it maximizes the correct detection probability for a bounded false alarm rate. To overcome the stability problems of previous approaches, a novel linear, parsimonious and non data-driven traffic model is proposed. This model remains stable in time and renders the process of traffic demand observable from SNMP measurements. The model can be used in two ways, either to estimate the anomaly-free OD flow volumes or to eliminate the anomaly-free traffic from the SNMP measurements in order to provide residuals sensitive to anomalies. Since a few anomaly-free SNMP measurements (at most one hour of measurements) is sufficient to obtain a reliable model of the OD flows, the proposed method is well adapted to highly non-stationary in time traffic and to dynamic routing. Using real traffic data from the Internet2 Abilene backbone network [32], we present an empirical comparison between our anomaly detection algorithm and the well known Principal Components Analysis (PCA) method introduced in [18]. The PCA approach has an important relevance in the anomaly detection field [18, 20, 22] but presents some important conception problems that we detect and analyze in our study. Through this analysis we verify the optimality properties of our detection algorithm and the stability of our traffic model, and show how our method outperforms the PCA approach in the considered dataset.

The remainder of this paper is organized as follows. The linear parsimonious OD flow model is introduced and validated in section 2. Section 3 describes the two different algorithms for anomaly detection that we compare in this work: our optimal detection algorithm and the previously introduced PCA approach. The evaluation and validation of our algorithm as well as a deep analysis of the PCA approach performance over real traffic data is conducted in section 4. Finally, section 5 concludes this work.

2   Handling Abrupt Traffic Changes

The anomaly detection algorithm that we present in this work consists of a non-sequential method. This algorithm presents optimality properties in terms of maximization of the detection probability for a bounded false alarm rate. To avoid direct OD flow measurements, the algorithm uses SNMP measurements $\mathbf{y}_t = \{y_t(1), \ldots, y_t(r)\}$ as input data; $y_t(i)$ represents the traffic volume (i.e. the amount of traffic) at link i in time interval t. High hardware requirements are necessary to network-wide collect and process direct OD flow measurements [7], so traffic models are generally developed using link SNMP measurements $\mathbf{y}_t$ and a routing matrix R to "reconstruct" OD flows. This reconstruction represents an ill-posed problem, as the number of unknown OD flows is much larger than the number of links [7]; in other words, it is not possible to directly retrieve the traffic demands $\mathbf{d}_t = \{d_t(1), \ldots, d_t(m)\}$ from $\mathbf{y}_t = R.\mathbf{d}_t$ given the ill-posed nature of the observation problem: $r << m$. Each traffic demand $d_t(i)$ represents the amount of traffic for OD couple i at time t. To overcome this difficulty, a parsimonious linear model for anomaly-free traffic is proposed. The idea of this model is that the anomaly-free traffic $\mathbf{d}_t$, sorted by OD flow volume can be decomposed at each time t over a known family of q basis functions $S = \{\mathbf{s}(1), \mathbf{s}(2), \ldots, \mathbf{s}(q)\}$ such that $q << m$. Therefore, the anomaly-free traffic can be expressed as $\mathbf{d}_t \approx S\boldsymbol{\mu}_t$ where the $m \times q$

matrix S is assumed to be known and $\boldsymbol{\mu}_t \in \mathbb{R}^q$ is a vector of unknown coefficients which describes the OD flows decomposition w.r.t. the set of vectors $\mathbf{s}(i)$. In this work, the traffic model is used to treat the anomaly-free traffic as a nuisance parameter, performing the anomaly detection in the traffic "residuals" that are obtained after removing the anomaly-free traffic. The anomaly-free traffic is removed by projection of the measured traffic on some space which is orthogonal to the space generated by the basis S. This transformation is based on the theory of invariance in statistics.

The parsimonious linear traffic model can be used to solve other problems than the anomaly detection one: TM estimation, using a least mean squares approach as it is shown in section 2.3, filtering and prediction with a Kalman approach, etc.

2.1   Stochastic Traffic Model for Anomaly Detection

It is assumed that the stochastic process of the anomaly-free OD traffic demand $\mathbf{d}_t$ obeys the following linear expression:

$$\mathbf{d}_t = \boldsymbol{\lambda}_t + \boldsymbol{\xi}_t \tag{1}$$

where $\boldsymbol{\lambda}_t \in \mathbb{R}^m$ is the mean traffic demand and $\boldsymbol{\xi}_t$ is a white Gaussian noise with covariance matrix $\Sigma = \text{diag}(\sigma_1^2, \ldots, \sigma_m^2)$. The process $\boldsymbol{\lambda}_t$ represents the "regular" part of the OD TM which can be correctly modeled when the behavior of the network is anomaly-free. The white Gaussian noise $\boldsymbol{\xi}_t$ models the natural variability of the OD TM together with the modeling errors. In order to describe the anomaly-free traffic $\boldsymbol{\lambda}_t$ with a small number of coefficients, a key feature of the TM is employed: its spatial stationarity; many classical TM models make use of this assumption, e.g. the gravity model [3,4,6]. The other key observation for this model is the "mice and elephants phenomenon": a small percentage of OD flows contribute to a large proportion of the total traffic [3,2]. The existence of such dominant flows together with the spatial stationarity of flows makes it reasonable to assume that, in the absence of an anomaly, the largest OD flows in a network remain the largest and the smallest flows remain the smallest during long periods of time; this assumption is confirmed in the empirical validation of the model, at least for several days, see section 2.3. Therefore, regarding the order of increasing OD flows (w.r.t. their traffic volume), it seems quite logical to accept that this order remains stable in time. It should be clear to the reader that this assumption can not be generalized to all network topologies and scenarios, but that holds for networks with a high level of aggregation (e.g. a backbone network or a large international VPN). The sorted OD flows can be interpreted as a discrete non-decreasing signal with certain smoothness. The curve obtained by interpolating this discrete signal is assumed to be a continuous curve, hence it can be parameterized by using a polynomial splines approximation.

Figure 2 shows the anomaly-free OD flows for the Abilene network, sorted in the increasing order of their volume of traffic, for different time instants t. The full lines depict the value of each sorted OD flow $d_t(k), k = 1..m$, the dashed

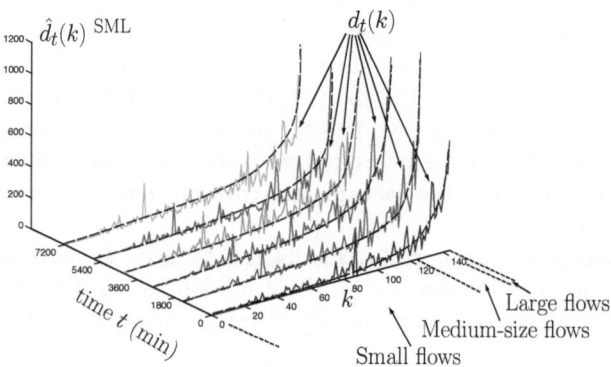

Fig. 2. Approximation of OD flows (full lines) by the spline-based model (dashed lines) for the Abilene network

lines represent the polynomial approximation of the sorted flows. In order to appreciate the time stability of this approximation, the curves are plotted for 6 consecutive days (from Sunday to Friday). Given the shape of the curve formed by the sorted OD flows, a cubic splines' approximation is applied; basic definitions and results on polynomial splines can be found in [27]. A discrete spline basis is designed, discretizing the continuous splines according to m points uniformly chosen in the interval $[1; m]$ and rearranging them according to the OD flows sorting order. The obtained linear parsimonious model for the anomaly-free traffic demand can be expressed as:

$$\mathbf{d}_t = S\boldsymbol{\mu}_t + \boldsymbol{\xi}_t \tag{2}$$

where $S = \{\mathbf{s}(i), i = 1..q\}$ is a $m \times q$ known matrix with a small number of columns w.r.t. m ($q << m$). The vectors $\mathbf{s}(i)$, which correspond to the rearranged discrete splines, form a set of known basis vectors describing the spatial distribution of the traffic; $\boldsymbol{\mu}_t = \{\mu_t(1)\ldots\mu_t(q)\}^T$ is the unknown time varying parameter vector which describes the OD flow intensity distribution with respect to the set of vectors $\mathbf{s}(i)$. The model for the anomaly-free link traffic is given by:

$$\mathbf{y}_t = G\boldsymbol{\mu}_t + \boldsymbol{\zeta}_t, \tag{3}$$

where $G = RS$ and $\boldsymbol{\zeta}_t \sim \mathcal{N}(0, \Phi)$, with $\Phi = R\Sigma R^T$. The computation of the rank of G is not simple since it depends on the routing matrix R. In practice, since the number of columns of G is very small, the product RS and its rank can be computed very fast. Therefore, it will be assumed that G is full column rank. To simplify notation and computations, the whitened measurements vector is introduced:

$$\mathbf{z}_t = \Phi^{-\frac{1}{2}}\mathbf{y}_t = H\boldsymbol{\mu}_t + \boldsymbol{\varsigma}_t, \tag{4}$$

where $H = \Phi^{-\frac{1}{2}} G$ and $\boldsymbol{\varsigma}_t \sim \mathcal{N}(0, I_r)$ (I_r is the $r \times r$ identity matrix). The purpose of this transformation is simply to whiten the Gaussian noise. Finally, the covariance matrix Σ is unknown. The solution consists of computing an estimate $\widehat{\Sigma}$ from a few anomaly-free measurements. Results on the estimation of $\widehat{\Sigma}$ can be found in [28].

2.2   Validation of the Model - The Dataset

The validation of the proposed traffic model is conducted using real data from the Abilene network, an Internet2 backbone network. Abilene consists of 12 router-level nodes and 30 OC192 links (2 OC48). The used router-level network topology and traffic demands are available at [33]. Traffic data consists of 6-months traffic matrices collected via Netflow from the Abilene Observatory [32]. The Abilene network is mainly a research experimental network; for this reason and as a particular case, the available dataset [33] consists of complete direct OD flow measurements $\mathbf{d}_t$. In order to reduce the overhead introduced by the direct measurement and process of flow-level data, our traffic model relies on SNMP links' load measurements $\mathbf{y}_t$. For the purpose of validation, we use the Abilene routing matrix R_o (available at [33]) to retrieve $\mathbf{y}_t$ from the OD flow measurements: $\mathbf{y}_t = R_o.\mathbf{d}_t$. In the following evaluations, we assume that traffic demands $\mathbf{d}_t$ are unknown and just consider the link load values $\mathbf{y}_t$ as the input known data.

The number of links is $r = 30$ and the number of OD flows is $m = 144$. The sampling rate is one measurement each 10 minutes. In order to verify the stability properties of the model, two sets of measurements are used: the first one, the "learning" anomaly-free dataset, is composed of one hour of anomaly-free SNMP measurements and it is used to construct the spline basis S; the second one, the "testing" dataset, is composed of 720 SNMP measurements (five days measurement period) and it is used to validate the model. Let T_{learning} (T_{testing} respectively) be the set of time indexes associated with SNMP measurements from the learning anomaly-free dataset (testing dataset respectively). The learning anomaly-free dataset is measured one hour before the testing dataset.

The same dataset is further used for the evaluation of the anomaly detection algorithms; therefore, the set of "true" anomalies is manually identified in the testing dataset. Manual inspection declares an anomaly in an OD flow if the unusual deviation intensity of the guilty OD flow leads to an increase of traffic (i) larger than 1.5% of the total amount of traffic on the network and (ii) larger than 1% of the amount of traffic carried by the links routing this guilty OD flow, for each of these links. Hence, only significant volume anomalies are considered as "true anomalies" (small volume anomalies have little influence on link utilization). Let $T_{\text{testing}}^{\text{free}} \subset T_{\text{testing}}$ be the set of time indexes associated with the 680 non-consecutive SNMP measurements of the testing dataset manually declared as anomaly-free (40 measurements of the testing dataset are affected by at least one significant volume anomaly).

2.3   Numerical Validation of the Model

Although many aspects could potentially be included in the evaluation, the size of the estimation error is considered as the quality indicator, using the root mean squared error (RMSE) as a measure of this size:

$$\text{RMSE}^{\text{label}}(t) = \sqrt{\sum_{k=1}^{m} \left(\hat{d}_t^{\text{label}}(k) - d_t(k) \right)^2}, \ \forall t \in T_{\text{testing}}^{\text{free}} \tag{5}$$

where $d_t(k)$ is the true traffic volume of the anomaly-free OD flow k at time t and $\hat{d}_t^{\text{label}}(k)$ denotes the corresponding estimate for the method entitled 'label'. Three estimates are compared: (i) simple gravity estimate [5] with label 'SG', (ii) tomogravity estimate [4,5] with label 'TG' and (iii) spline-based Maximum Likelihood (ML) estimate with the label 'SML'. Since the traffic linear model is a Gaussian model, the Maximum Likelihood estimate of $\mathbf{d}_t$, namely $\hat{\mathbf{d}}_t^{\text{SML}}$ corresponds to the least mean squares estimate, given by $\hat{\mathbf{d}}_t^{\text{SML}} = S(H^T H)^{-1} H^T \mathbf{z}_t$. The statistical properties of the ML estimate are well known [28] contrary to the simple gravity and tomogravity estimates. The spline-based model is computed using the learning dataset, following these steps: (i) the tomogravity estimate $\hat{d}_t^{\text{TG}}(k)$ is computed for all OD flows k and all $t \in T_{\text{learning}}$, (ii) the mean flow values $\bar{d}^{\,\text{TG}}(k) = \frac{1}{\text{card}(T_{\text{learning}})} \sum_{t \in T_{\text{learning}}} \hat{d}_t^{\text{TG}}(k)$ are computed, where $\text{card}(T_{\text{learning}})$ is the number of time indexes in the learning dataset and (iii) sorted in ascending order to obtain a rough estimate of the OD flows traffic volume. The spline-based model is designed with cubic splines and 2 knots (representing small, medium-size and large OD flows). The mean value $\bar{d}^{\,TG}(k)$ is also used to compute an estimate $\hat{\sigma}_k^2$ of σ_k^2, which leads to an estimate $\hat{\Phi}$ of Φ.

Figure 3 depicts the error $\text{RMSE}^{\text{label}}(t)$ over the set $T_{\text{testing}}^{\text{free}}$. The total error in $T_{\text{testing}}^{\text{free}}$, $\text{TRMSE}^{\text{label}} = \sum_{t \in T_{\text{testing}}^{\text{free}}} \text{RMSE}^{\text{label}}(t)$ is presented in table 1 as a global indicator of methods' performance. The spline-based estimate outperforms the other estimates, as it produces the smallest total estimation error. The TRMSE corresponding to the tomogravity estimate TG is quite close to the error produced with our model. However, the SML estimate presents a major advantage w.r.t. the TG estimate: as it was previously said, the ML estimate presents

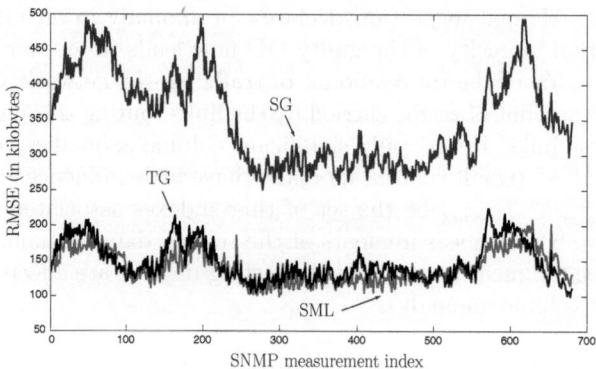

Fig. 3. Comparison between the SG, TG and SML RMSE for 680 anomaly-free measurements

Table 1. TRMSE (in kilobytes) for 680 anomaly-free measurements for gravity (SG), tomogravity (TG) and spline-based (SML) models

Method	SG	TG	SML
TRMSE (kB)	9337	3935	3766

well established statistical properties, which is not the case for the TG estimate. The SML estimate is asymptotically optimal, i.e. it is asymptotically unbiased and efficient. Moreover, the spline-based model can be used in order to design anomaly detection algorithms with optimality properties, which is not the case for the tomogravity estimate. As a final validation, the Gaussian assumption of the model is studied. The "residuals" of measurements are analyzed, i.e. the obtained traffic after filtering the "regular" part, $G\mu_t$. The residuals are obtained by projection of the whitened measurements vector $\mathbf{z}_t = \Phi^{-\frac{1}{2}}\mathbf{y}_t$ onto the left null space of H, using a linear transformation into a set of $r - q$ linearly independent variables $\mathbf{u}_t = W\mathbf{z}_t \sim \mathcal{N}(0, I_{r-q})$. The matrix W is the linear rejector that eliminates the anomaly-free traffic, built from the first $r - q$ eigenvectors of the projection matrix $P_H^\perp = I_r - H(H^T H)^{-1} H^T$ corresponding to eigenvalue 1.

The rejector verifies the following relations: $WH = 0$, $W^T W = P_H^\perp$ and $WW^T = I_{r-q}$. $P_H^\perp$ represents the projection matrix onto the left null space of H. The Kolmogorov-Smirnov test [29] at the level 5% accepts the Gaussian hypothesis for 670 of the 680 measurements with time indexes in $T_{\text{testing}}^{\text{free}}$ (acceptation ratio of 98.5%), which confirms the Gaussian assumption.

3   Optimal Anomaly Detection and the PCA Approach

In this section we introduce the optimal volume anomaly detection algorithm. The goal of the proposed method is to detect an additive change θ in one or more OD flows of the traffic demand time series $\mathbf{d}_t$ from a sequence of SNMP measurements $\mathbf{y}_t = R.\mathbf{d}_t$. For technical reasons, it will be assumed that the amplitude of the change θ is constant; however, as it is shown in the results from section 4, this technical assumption does not restrict in practice the applicability of the proposed approach. Our detection algorithm consists of a non-sequential approach, also known as a "snapshot" method. Non-sequential approaches allow to define optimal algorithms, regarding the maximization of the probability of anomaly detection and the minimization of false alarms (i.e. raising an alarm in the absence of an anomaly). In this work, a simple snapshot approach is presented, which allows to detect an anomaly with the highest probability of detection for a given probability of false alarm. The previously introduced anomaly-free traffic model is slightly modified, in order to explicitly consider the temporal variation of the covariance matrix Σ. The Gaussian noise ξ_t is now assumed to have a covariance matrix $\gamma_t^2 \Sigma$; $\Sigma = \text{diag}(\sigma_1^2, \ldots, \sigma_m^2)$ is assumed to be known and stable in time. The scalar γ_t is unknown and serves to model the mean level of

OD flows' volume variance. As it is explained in [28], this distinction between Σ and γ_t was not necessary in section 3. However, in the detection problem, this separation allows to accurately define the detection thresholds.

3.1   Optimal Volume Anomaly Detection

Typically, when an anomaly occurs in one or several OD flows, the measurement vector **y** presents an abrupt change in those links where the OD flows are routed. The detection of this anomalous increase can be treated as a hypothesis testing problem, considering two alternatives: the null hypothesis $\mathcal{H}_0$, where OD flows are anomaly-free and the alternative hypothesis $\mathcal{H}_1$, where OD flows present an anomaly:

$$\mathcal{H}_0 = \{\mathbf{z} \sim \mathcal{N}(\varphi + H\boldsymbol{\mu}, \gamma_t^2 I_r); \; \varphi = 0, \; \boldsymbol{\mu} \in \mathbb{R}^q\}, \tag{6}$$

$$\mathcal{H}_1 = \{\mathbf{z} \sim \mathcal{N}(\varphi + H\boldsymbol{\mu}, \gamma_t^2 I_r); \; \varphi \neq 0, \; \boldsymbol{\mu} \in \mathbb{R}^q\}. \tag{7}$$

Here φ represents the evidence of an anomaly. In the anomaly detection problem, $\boldsymbol{\mu}$ is considered as a nuisance parameter since (i) it is completely unknown, (ii) it is not necessary for the detection and (iii) it can mask the anomalies. It is possible to decide between $\mathcal{H}_0$ and $\mathcal{H}_1$ if, in the case of an anomaly, φ has a non-null component in the left null space of H. This verifies for any value of φ in the form of $\varphi = \theta \Phi^{-\frac{1}{2}} \mathbf{r}^*$, where $\mathbf{r}^*$ stands for the sum of the normalized columns of the routing matrix R with indexes corresponding to the anomalous demands.

The quality of a statistical test is defined by the false alarm rate and the power of the test. The above mentioned testing problem is difficult because (i) $\mathcal{H}_0$ and $\mathcal{H}_1$ are composite hypotheses and (ii) there is an unknown nuisance parameter $\boldsymbol{\mu}$. There is no general way to test between composite hypotheses with a nuisance parameter. In this paper we use the statistical test $\phi^* : \mathbb{R}^r \mapsto \{\mathcal{H}_0, \mathcal{H}_1\}$ of [31], inspired by the fundamental paper of Wald [30]. The test is designed as:

$$\phi^*(\mathbf{z}) = \begin{cases} \mathcal{H}_0 & \text{if } \Lambda(\mathbf{z}) = \|P_H^\perp \mathbf{z}\|^2 / \gamma_t^2 < \lambda_\alpha \\ \mathcal{H}_1 & \text{else} \end{cases} \tag{8}$$

where $\| \cdot \|$ represents the Euclidean norm.

Let K_α be the class of tests with an upper bounded maximum false alarm probability, $K_\alpha = \{\phi : \sup_{\boldsymbol{\mu}} \Pr_{\varphi=0,\boldsymbol{\mu}}(\phi(\mathbf{z}) = H_1) \leqslant \alpha\}, 0 < \alpha < 1$; here $\Pr_{\varphi=0,\boldsymbol{\mu}}$ stands for the probability when $\mathbf{z} \sim \mathcal{N}(H\boldsymbol{\mu}, \gamma_t^2 I_r)$. The power function or hit rate is defined as $\beta_\phi(\varphi, \boldsymbol{\mu}) = \Pr_{\varphi \neq 0, \boldsymbol{\mu}} (\phi(\mathbf{z}) = \mathcal{H}_1)$. A priori, this probability depends on the nuisance parameter $\boldsymbol{\mu}$ as well as on the parameter φ which is highly undesirable. However, the test $\phi^*(\mathbf{z})$ defined by equation (8) has uniformly best constant power (UBCP) in the class K_α over the family of surfaces $S = \{S_c : c \geq 0\}$ defined by $S_c = \{\varphi : \|P_H^\perp \varphi\|^2 = c^2\}$. UBCP means that $\beta_{\phi^*}(\varphi, \boldsymbol{\mu}) = \beta_{\phi^*}(\varphi', \boldsymbol{\mu}), \forall \varphi, \varphi' \in S_c$ and $\beta_{\phi^*}(\varphi, \boldsymbol{\mu}) \geqslant \beta_\phi(\varphi, \boldsymbol{\mu})$ for any test $\phi \in K_\alpha$. The threshold λ_α is chosen to satisfy the false alarm bound α, $\Pr_{\varphi=0,\boldsymbol{\mu}}(\Lambda(\mathbf{z}) \geqslant \lambda_\alpha) = \alpha$. The UBCP property of this test represents the optimality condition of the detection algorithm.

3.2 Principal Components Analysis for Anomaly Detection

The Principal Components Analysis (PCA) approach for anomaly detection [18, 22, 20] consists of a two steps methodology: (i) parsimonious anomaly-free traffic modeling, using a decomposition of traffic measurements into a principal components basis and (ii) anomaly detection in the traffic residuals, i.e. the traffic not described by the PCA decomposition. PCA is a linear coordinate transformation that maps a given set of data points to a new coordinate system, such that the greatest variance of any projection lies on the first coordinate $\mathbf{w}_1$ (called the first principal component or first PC), the second greatest variance on the second coordinate $\mathbf{w}_2$, and so on. Given a traffic measurement matrix $\mathbf{Y} \in \mathbb{R}^{p \times r}$, where each column represents a time series of p samples of SNMP measurements for each link, the PCA traffic modeling consists of computing the r principal components of $\mathbf{Y}$, $\mathbf{w}_{i=1..r}$, using the first k principal components to capture the anomaly-free behavior of traffic and the remaining $r - k$ components to construct residuals sensitive to anomalies. The first k principal components are the "normal components" and the remaining $r - k$ are the "anomalous components". Each of the principal components can be computed as follows:

$$\mathbf{w}_1 = \arg \max_{||\mathbf{w}||=1} ||\mathbf{Y}\mathbf{w}||$$

$$\mathbf{w}_k = \arg \max_{||\mathbf{w}||=1} ||(\mathbf{Y} - \sum_{i=1}^{k-1} \mathbf{Y}\mathbf{w}_i\mathbf{w}_i^T)\mathbf{w}||$$

The idea behind this approach is that traffic anomalies are sparse in $\mathbf{Y}$, and so the first components of the transformation will correctly describe the anomaly-free behavior. The space spanned by the set of normal components is the "normal subspace" S and the space spanned by the anomalous components is the "anomalous sub-space" $\hat{S}$. After the construction of the normal and anomalous sub-spaces, the links' traffic $\mathbf{y}$ can be separated at each time t in the modeled traffic $\mathbf{y}_{\text{model}}$ and the residual traffic $\mathbf{y}_{\text{residual}}$ by simple projection onto S and $\hat{S}$:

$$\mathbf{y} = \mathbf{y}_{\text{model}} + \mathbf{y}_{\text{residual}}$$

$$\mathbf{y}_{\text{model}} = \mathbf{PP}^T\mathbf{y}$$

$$\mathbf{y}_{\text{residual}} = (\mathbf{I} - \mathbf{PP}^T)\mathbf{y}$$

where $\mathbf{P} \in \mathbb{R}^{r \times k}$ stands for the matrix with the first k PCs as column vectors and $\mathbf{PP}^T$ represents the projection matrix onto the normal sub-space. The anomaly detection is then performed in the residual traffic, looking for large changes in the squared norm of residuals, $||\mathbf{y}_{\text{residual}}||^2$.

4 Validation of the Detection Algorithm and PCA Evaluation

The detection algorithm is applied to the SNMP measurements of the testing dataset. The false alarm probability is fixed to $\alpha = 0.01$. For the detection

Table 2. Results of the detection for 720 measurements composed of 680 anomaly-free measurements and 40 anomalous measurements for the spline-based and PCA tests

Situation	Spline-based	PCA (1 PC)
Normal operation	672 (98.82 %)	671 (98.68 %)
False alarms	8 (1.18 %)	9 (1.32 %)
Missed detections	9 (22.50 %)	25 (62.50 %)
Correct detections	31 (77.50 %)	15 (37.50 %)

purpose, it is crucially important to have a good estimate of γ_t. This parameter is estimated from the learning dataset by using the ML estimate of noise variance [28] in residuals $\mathbf{u}_t$. Since this parameter can slowly vary in time, its value is updated during the test: at time t, if no anomaly has been declared in the last hour, γ_t is estimated by its value one hour before. The performance of our method is compared to the performance obtained with the PCA approach. This method is chosen as benchmark given its relevance in the anomaly detection literature [22,18,20]. The obtained results are presented in table 2. The column *Spline-based* shows that the proposed test (8) obtains a false alarm rate of 1.18%, close to the prescribed value $\alpha = 0.01$. The probability to detect a volume anomaly is about 77.5%. The column *PCA* presents the results obtained with the PCA approach. The best performance that can be attained with the PCA test is considered in this evaluation, using just the first PC to model the normal sub-space; the following discussion about results in figure 4 clarifies this election. The detection threshold of this test is chosen to obtain a similar false alarm rate of 1.32%. The PCA test presents a very low correct detection rate for this level of false alarm, about 37.50%. Figure 4 illustrates the ROC curves for the Spline-based and the PCA tests for different number of first PCs to model the normal sub-space. The figure presents the correct detection rate β for different values of the false alarm rate α. The ROC curves allow to compare the accuracy of both tests and the sensitivity of each detection method w.r.t. the variation of the detection thresholds, showing the existing trade-off between the correct detection and the false alarm rates. Results obtained with the PCA approach in the Abilene dataset are far from those obtained with our method; the PCA test presents more than 2 times lower detection rates for a reasonable false alarm rate, below 5%. There are at least three major problems regarding the PCA approach: (i) its performance strongly depends on the number of components selected to describe the normal space; (ii) the traffic modeling procedure is data-driven, posing serious stability problems and (iii) the learning step is unsupervised but very time-consuming, becoming prone to bad-learning effects. Similar problems were also analyzed and verified by the authors of the original PCA for anomaly detection approach [18] in [21, 22]. Let us begin by the first issue; in [18], the separation between the normal and anomalous principal components is performed using a simple ad-hoc threshold-based separation method that is highly tuned for each dataset and cannot therefore be generalized, making the PCA approach inapplicable in a general scenario. Figure 5 depicts the temporal evolution of $||\mathbf{y}_{\text{residual}}||^2$,

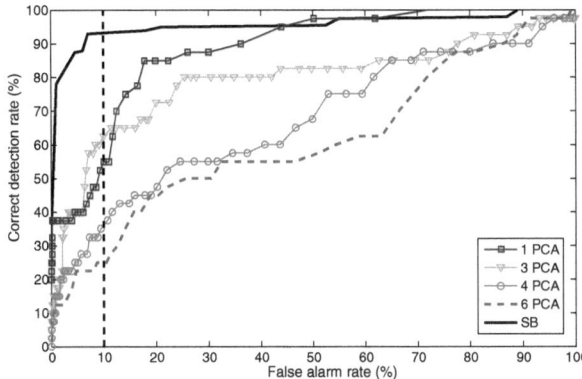

Fig. 4. Correct detection rate vs false alarm rate for the spline-based test (SB - solid line) and the PCA test, considering different number of first PCs to model the normal sub-space

using a different number of PCs to describe the normal sub-space (1, 2, 4 and 5 first PCs are used to model the anomaly-free traffic). The dotted line represents the detection threshold; the squares indicate the times when an anomaly truly occurs, according to the manual inspection performed in section 2.2. It can be appreciated that the false positive rate is very sensitive to small differences in the number of principal components used to describe the normal sub-space. The ROC curves in figure 4 show that there is no single PCA representation for the Abilene dataset that offers a good balance between correct detection and false alarm rates.

Regarding the second issue, the traffic modeling in the PCA approach is data-driven, i.e. the PCA decomposition strongly depends on the considered SNMP measurements matrix $\mathbf{Y}$. In [18], the normal and anomalous sub-spaces are constructed from a given matrix $\mathbf{Y}^o$ at a certain time window t_o, and the representation is assumed to be stable during long-time periods, from week to week. However, it is easy to see that this approach is highly unstable, even from one time window to the other. Let us consider an extreme-case example that will also illustrate the learning problems of the approach. The PCA approach assumes that the normal sub-space can be correctly described by the first principal components of $\mathbf{Y}^o$ as they capture the highest level of "energy". Figure 6 depicts the temporal evolution of the variance captured by each principal component $\mathbf{w}_i$, $\|\mathbf{Y}\mathbf{w}_i\|^2$, considering time windows of 12hs (i.e. the set of PCs is recomputed every 12hs). In almost every time window, the first principal component captures the highest energy, justifying the use of one single PC to describe the normal traffic behavior. However, large anomalies at time windows t_3 and t_8, also visible in figure 5.(a) contribute to a large proportion of the captured energy; in this case, a second principal component may be added as a descriptor of the normal traffic. Since this second component corresponds in fact to an anomaly, the normal sub-space is inadvertently polluted, turning useless the learning step.

110   P. Casas et al.

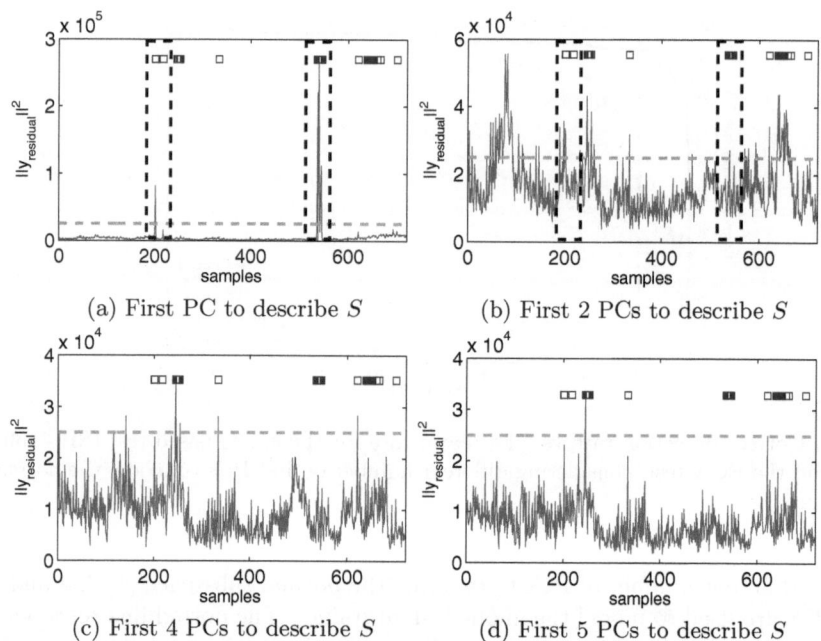

(a) First PC to describe S (b) First 2 PCs to describe S

(c) First 4 PCs to describe S (d) First 5 PCs to describe S

Fig. 5. Temporal evolution of $||y_{\mathrm{residual}}||^2$, using a different number of first PCs to model the normal sub-space S. The squares indicate when an anomaly truly occurs. The dotted line depicts the detection threshold. Large anomalies pollute the normal sub-space and are not detected with the PCA approach. (a) Both large anomalies at samples 200 and 540 are correctly detected using 1 PC to describe S. (b) Large anomalies are not detected using a 2 PCs representation of S.

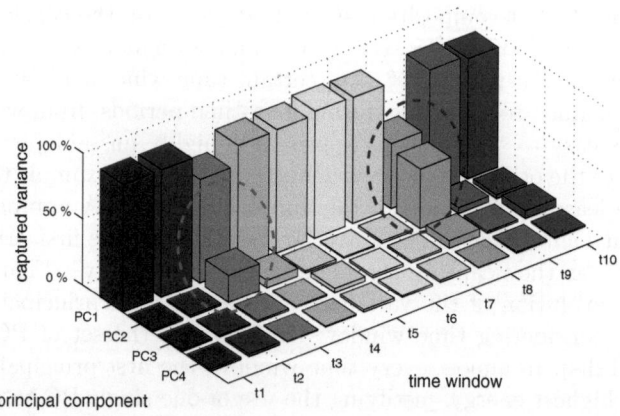

Fig. 6. Temporal evolution of the total variance captured by each PC $\mathbf{w}_i$, $||\mathbf{Y}\mathbf{w}_i||^2$. Each time window $t_{j=1..10}$ consists of 12hs of SNMP data. Large anomalies may inadvertently pollute the normal sub-space at t_3 and t_8.

In figure 5.(b), both large anomalies at t_3 and t_8 are not detected due to this effect.

This brings us to the last but not least problem; the learning step of the PCA approach is very "time-consuming": the number of samples p must be greater than the number of links r, in order to obtain at least r independent PCs [22]. In this sense, the approach is more prone to suffer from this kind of polluting effect, since it is likely that an anomaly occurs on longer time periods. Our algorithm is not data-driven and has a very short learning-step: as we show in section 2.3, at most one hour of measurements is sufficient to obtain a reliable model of the OD flows. The effect of a training step over polluted data does not represent a problem to our short-learning approach, as it is quite simple to assure or look for a 1-hour anomaly-free time period.

5   Conclusions and Some Extensions

In this paper, we have presented and evaluated a new statistical algorithm for volume anomaly detection in data networks. This algorithm presents well-established optimality properties in terms of detection probability and false alarm generation, unavailable in previous proposals in the field and extremely important in order to provide solid results. For the purpose of anomaly detection, we have introduced an original linear parsimonious spline-based traffic model which allows to treat the anomaly-free traffic as a nuisance parameter. This model parameterizes traffic flows from simple link load measurements, reducing the overhead of direct flow measurements. Compared to other different traffic models, this model is not data-driven and remains stable in time, a necessary property to achieve reliable results. We have also applied this traffic model to the traffic matrix estimation problem, achieving better results than those obtained with classical models (e.g. tomogravity model). We have analyzed the performance of a very well known anomaly detection method, the so called PCA for anomaly detection approach into a real traffic dataset and studied in depth some of the weaknesses of this approach. We have finally compared our algorithm to the PCA approach and showed that the spline-based anomaly detection test outperforms the PCA based approach as predicted by the optimality properties of the test, which highlights the impact of our proposal for volume anomaly detection. In this work we have only treated the anomaly detection problem. In [1] we present some interesting countermeasures to react against anomalies, based on routing reconfiguration.

References

1. Casas, P., Fillatre, L., Vaton, S.: Robust and Reactive Traffic Engineering for Dynamic Traffic Demands. In: Proc. EuroNGI Conference on Next Generation Networks (2008)
2. Johansson, C., Gunnar, A.: Data-driven Traffic Engineering: techniques, experiences and challenges. In: Proc. IEEE BROADNETS (2006)

3. Medina, A., Salamatian, K., Bhattacharyya, S., Diot, C.: Traffic Matrix Estimation: Existing Techniques and New Directions. In: Proc. ACM SIGCOMM (2002)
4. Zhang, Y., Roughan, M., Lund, C., Donoho, D.: Estimating Point-to-Point and Point-to-Multipoint Traffic Matrices: an Information-Theoretic Approach. IEEE/ACM Trans. Networking 13(5), 947–960 (2005)
5. Zhang, Y., Roughan, M., Duffield, N., Greenberg, A.: Fast Accurate Computation of Large-Scale IP Traffic Matrices from Link Load Measurements. In: Proc. ACM SIGMETRICS (2003)
6. Gunnar, A.,, Johansson, M., Telkamp, T.: Traffic Matrix Estimation on a Large IP Backbone - A Comparison on Real Data. In: Proc. USENIX/ACM IMC (2004)
7. Coates, M., Hero, A., Nowak, R., Yu, B.: Internet Tomography. IEEE Signal Processing Magazine 19(3), 47–65 (2002)
8. Hood, C., Ji, C.: Proactive network fault detection. In: Proc. IEEE INFOCOM (1997)
9. Katzela, I., Schwartz, M.: Schemes for fault identification in communications networks. IEEE/ACM Trans. Networking 3(6), 753–764 (1995)
10. Ward, A., Glynn, P., Richardson, K.: Internet service performance failure detection. Performance Evaluation Review (1998)
11. Jung, J., Krishnamurthy, B., Rabinovich, M.: Flash crowds and denial of service attacks: Characterization and implications for CDNs and webs. In: Proc. ACM WWW 2002 (2002)
12. Xie, L., et al.: From Detection to Remediation: A Self-Organized System for Addressing Flash Crowd Problems. In: Proc. IEEE ICC (2008)
13. Barford, P., Kline, J., Plonka, D., Ron, A.: A Signal Analysis of Network Traffic Anomalies. In: ACM SIGCOMM Internet Measurement Workshop (2002)
14. Brutlag, J.D.: Aberrant Behavior Detection in Time Series for Network Monitoring. In: Proc. 14th Systems Administration Conference (2000)
15. Cheng, C.M., Kung, H., Tan, K.S.: Use of Spectral Analysis in Defense Against DoS Attacks. In: Proc. IEEE GLOBECOM (2002)
16. Zou, C.C., Gong, W., Towsley, D., Gao, L.: The Monitoring and Early Detection of Internet Worms. IEEE/ACM Trans. Networking 13(5), 961–974 (2005)
17. Wang, H., Zhang, D., Shin, K.: Detecting SYN flooding attacks. In: Proc. IEEE INFOCOM (2002)
18. Lakhina, A., Crovella, M., Diot, C.: Diagnosing Network-Wide Traffic Anomalies. In: Proc. ACM SIGCOMM (2004)
19. Lakhina, A., Crovella, M., Diot, C.: Mining Anomalies Using Traffic Feature Distributions. In: Proc. ACM SIGCOMM (2005)
20. Li, X., Bian, F., Crovella, M., Diot, C., Govindan, R., Iannaccone, G., Lakhina, A.: Detection and Identification of Network Anomalies Using Sketch Subspaces. In: Proc. USENIX/ACM IMC (2006)
21. Ahmed, T., Coates, M., Lakhina, A.: Multivariate Online Anomaly Detection Using Kernel Recursive Least Squares. In: Proc. IEEE INFOCOM (2007)
22. Ringberg, H., Soule, A., Rexford, J., Diot, C.: Sensitivity of PCA for Traffic Anomaly Detection. In: Proc. ACM SIGMETRICS (2007)
23. Thottan, M., Ji, C.: Anomaly Detection in IP Networks. IEEE Trans. Signal Processing 51(8), 2191–2204 (2003)
24. Soule, A., Salamatian, K., Taft, N.: Combining Filtering and Statistical Methods for Anomaly Detection. In: Proc. USENIX/ACM IMC (2005)
25. Tartakovsky, A., et al.: A novel approach to detection of intrusions in computer networks via adaptive sequential and batch-sequential change-point detection methods. IEEE Trans. Signal Processing 54(9), 3372–3382 (2006)

26. Zhang, Y., Ge, Z., Greenberg, A., Roughan, M.: Network Anomography. In: Proc. USENIX/ACM IMC (2005)
27. Nürnberger, G.: Approximation by Spline Functions. Springer, Heidelberg (1989)
28. Rao, C.: Linear Statistical Inference and its Applications. J. Wiley & Sons, Chichester (1973)
29. Lehman, E.: Testing Statistical Hypotheses, 2nd edn. Chapman & Hall, Boca Raton (1986)
30. Wald, A.: Tests of statistical hypotheses concerning several parameters when the number of observations is large. Trans. American Math. Soc. 54, 426–482 (1943)
31. Fillatre, L., Nikiforov, I.: Non-bayesian detection and detectability of anomalies from a few noisy tomographic projections. IEEE Trans. Signal Processing 55(2), 401–413 (2007)
32. The Abilene Observatory, http://abilene.internet2.edu/observatory/
33. Zhang, Y.: Abilene Dataset 04, http://www.cs.utexas.edu/yzhang/

Traffic Engineering of Telecommunication Networks Based on Multiple Spanning Tree Routing

Dorabella Santos[1], Amaro de Sousa[2], and Filipe Alvelos[3]

[1] Instituto de Telecomunicações
3810-193 Aveiro, Portugal
dorabella@av.it.pt
[2] Instituto de Telecomunicações / DETI
Universidade de Aveiro
3810-193 Aveiro, Portugal
asou@ua.pt
[3] Centro Algoritmi / DPS
Universidade do Minho
4710-057 Braga, Portugal
falvelos@dps.uminho.pt

Abstract. This paper focuses on traffic engineering of telecommunication networks, which arises in the context of switched Ethernet networks. It addresses the minimization of the maximum network link load. With the IEEE 802.1s Multiple Spanning Tree Protocol, it is possible to define multiple routing spanning trees to provide multiple alternatives to route VLAN traffic demands. Two compact mixed integer linear programming models defining the optimization problem and several models based on the Dantzig-Wolfe decomposition principle, which are solved by branch-and-price, are proposed and compared. The different decompositions result from defining as subproblems either the supporting spanning trees and/or the demand routing paths, which can be solved by well known efficient algorithms.

Keywords: multiple spanning tree routing, integer programming, column generation.

1 Introduction

In this paper, we deal with the minimization of the maximum link load of telecommunication networks based on multiple spanning tree routing. This is an important traffic engineering objective not only because it maximizes network robustness to unpredicted growth of traffic demand but also because it decreases the amount of traffic that suffers service disruption and requires rerouting when network links fail. The multiple spanning tree routing paradigm arises in the context of switched Ethernet networks with the IEEE 802.1s Multiple Spanning Tree Protocol (MSTP) [1]. In these networks, traffic flows are defined on a per

R. Valadas and P. Salvador (Eds.): FITraMEn 2008, LNCS 5464, pp. 114–129, 2009.
© Springer-Verlag Berlin Heidelberg 2009

VLAN basis and routing of VLAN traffic flows is done through the network based on spanning trees (STs). The network operator defines each required ST by creating a spanning tree instance (STI) and assigning to it a BridgeID value per switch and a PortCost value per interface (STI parameters are independent between STIs). At all moments, MSTP sets the active links of each STI based on the minimum cost paths of every switch to the root bridge (which is the switch with the lowest BridgeID value). When a VLAN is assigned to a STI, its traffic flows are routed between its end nodes through the unique path defined by the assigned STI. In this paper, we consider E-Line VLANs [2], i.e., point-to-point VLANs with a single bi-directional commodity between two end switches.

Note that the STI parameters define the ST set of active links not only in the normal network operation but also in all failure situations. In this paper, we address the normal operation state and, in this case, it is possible to separate the STI parameter assignment from the routing optimization problem since it is always possible to compute a set of ST parameters for every required ST. Given a set of network links defining a spanning tree, an ST parameter assignment procedure is proposed in [3] that also ensures minimum service disruption, i.e., minimizes the number of VLAN flows that suffer service disruption for every single link failure.

MSTP also enables the set of STIs over a set of regions defined on the network. In this case, MSTP ensures that failures inside a region do not affect routing outside that region minimizing the impact of network failures in the overall network stability. In [4], the authors address the problem of how to divide the network into regions. Nevertheless, it has been shown in [3], [5] that, with optimized configurations, considering the whole network as a single region provides the best results with respect to load balance and service disruption and, here, we consider the single region approach. The use of MSTP as a means to enhance traffic engineering capabilities of Ethernet networks has been addressed by other authors [6], [7], [8], [9] and other works propose MSTP as a means to improve the support to other important aspects like mobility [10] and QoS [11]. Nevertheless, as far as we are aware, there is no reference dealing with exact methods to find optimal solutions for minimizing the maximum link load of multiple spanning tree routing based networks such as Ethernet.

In this paper, we propose two compact mixed integer linear programming models defining the minimization of the maximum link load applied to spanning tree routing networks: a first natural model and a second which is a reformulation of the first one and more efficient when solved by standard optimization packages (we have used CPLEX). We also describe several models based on the Dantzig-Wolfe decomposition principle, which can be solved by branch-and-price, as an alternative means to solve the optimization problem. The different decompositions result from defining as subproblems either the supporting spanning trees and/or the demand routing paths of the compact models, which can be solved by well known efficient algorithms. The computational results show that the efficiency of the different decompositions depends on the compact model that they are based on. Nevertheless, the decompositions are able to reduce the number of nodes of the branch-and-bound search when compared with CPLEX.

This paper is organized as follows. In section II, the two compact mixed integer linear programming models defining the optimization problem are proposed. Section III presents the different models based on the Dantzig-Wolfe decomposition principle applied to the previous compact models. Section IV presents and discusses the computational results obtained with the compact models and with the decompositions. At the end, some final conclusions are drawn in Section V.

2   Compact Models

Consider a network defined on a graph $G(N, A)$ where N is the set of nodes and A is the set of links between nodes. Edge $\{i, j\} \in A$ represents a link between node $i \in N$ and $j \in N$, while arc $(i, j) \in A$ represents edge $\{i, j\}$ directed from i to j. Each edge has a known capacity given by $c_{\{ij\}}$.

The network has to support a set of commodities K where each commodity $k \in K$ is characterized by its origin node $o_k \in N$, its destination node $d_k \in N$ and its demand $b_k \in \mathbb{R}^+$ (the demand is bi-directional, i.e., it is the same from o_k to d_k as from d_k to o_k).

Commodities must be routed over the network on paths defined by one of a set of spanning trees S of size $|S|$. The aim is to choose set S and to assign the commodities to the spanning trees in order to minimize the maximum network link load. To solve the optimization problems, $|S|$ should not be too large, or else the number of binary variables explodes. Nevertheless, it has been shown in [12] that a "good" number for $|S|$ is roughly the number of disjoint spanning trees that one can identify on graph $G(N, A)$, which is usually a small number in practice.

2.1   First Compact Model

Let $V(i)$ be the set of neighbor nodes of $i \in N$ on network $G(N, A)$ and r be one node chosen from set N and referred to as "root node" (it is used to model the spanning tree constraints).

In a node-link formulation, the basic decision variables are:

$\mu \in [0, 1]$ the maximum load of any edge
$x_{ij}^k \in \{0, 1\}$ indicates if arc $(i, j) \in A$ is in the path of the commodity $k \in K$
$\phi_k^s \in \{0, 1\}$ indicates if commodity $k \in K$ is assigned to spanning tree $s \in S$
$\beta_{\{ij\}}^s \in \{0, 1\}$ indicates if edge $\{i, j\} \in A$ is in the spanning tree $s \in S$.

Additionally, we need the following variables in order to obtain an appropriate model for the spanning tree constraints:

$y_{ij}^{zs} \in \{0, 1\}$ indicates if arc $(i, j) \in A$ is in the path from node $z \in N \backslash \{r\}$ to the
           root node r on the spanning tree $s \in S$
$\theta_{ij}^s \in \{0, 1\}$ indicates if arc $(i, j) \in A$ is in the spanning tree $s \in S$.

The Mixed Integer Linear Programming (MILP) model defining the optimization problem is given by:

$$\min \mu$$

s.t.

$$\sum_{j \in V(i)} \left(y_{ij}^{zs} - y_{ji}^{zs} \right) = \begin{cases} 1, i = z \\ 0, i \neq z \end{cases} \qquad \forall z \in N \backslash \{r\}, \forall s \in S, \forall i \in N \backslash \{r\} \tag{1}$$

$$\theta_{ij}^{s} \geq y_{ij}^{zs} \qquad \forall z \in N \backslash \{r\}, \forall (i,j) \in A, \forall s \in S \tag{2}$$

$$\sum_{j \in V(i)} \theta_{ij}^{s} = \begin{cases} 1, i \neq r \\ 0, i = r \end{cases} \qquad \forall i \in N, \forall s \in S \tag{3}$$

$$\theta_{ij}^{s} + \theta_{ji}^{s} = \beta_{\{ij\}}^{s} \qquad \forall \{i,j\} \in A, \forall s \in S \tag{4}$$

$$x_{ij}^{k} + x_{ji}^{k} \leq \beta_{\{ij\}}^{s} + (1 - \phi_{k}^{s}) \qquad \forall \{i,j\} \in A, \forall k \in K, \forall s \in S \tag{5}$$

$$\sum_{j \in V(i)} \left(x_{ij}^{k} - x_{ji}^{k} \right) = \begin{cases} 1, & i = o_k \\ 0, & i \neq o_k, d_k \ \forall i \in N, \forall k \in K \\ -1, i = d_k \end{cases} \tag{6}$$

$$\sum_{s \in S} \phi_{k}^{s} = 1 \qquad \forall k \in K \tag{7}$$

$$\sum_{k \in K} b_k \left(x_{ij}^{k} + x_{ji}^{k} \right) \leq c_{\{ij\}} \mu \qquad \forall \{i,j\} \in A \tag{8}$$

$$y_{ij}^{zs} \in \{0,1\}, \theta_{ij}^{s} \in \{0,1\}, \phi_{k}^{s} \in \{0,1\}, x_{ij}^{k} \in \{0,1\}, \beta_{\{ij\}}^{s} \in \{0,1\}, \mu \in [0,1] \ .$$

The optimization objective is to minimize the value of variable μ that accounts for the maximum load of any edge.

Constraints (1)–(4) guarantee that the set of variables $\beta_{\{ij\}}^{s}$ define the spanning trees. For each $s \in S$: constraints (1) guarantee that variables y_{ij}^{zs} define a path from $z \in N$ to root node $r \in N$; constraints (2) guarantee that variables θ_{ij}^{s} include the arcs of all paths defined by variables y_{ij}^{zs}; constraints (3), together with constraints (2), guarantee that the paths given by variables y_{ij}^{zs} define a directed spanning tree towards the root node r; constraints (4) guarantee that the variables $\beta_{\{ij\}}^{s}$ include the edges of the directed spanning tree defined by variables θ_{ij}^{s}.

Constraints (6) are the path conservation constraints which together with (5) guarantee that the path is defined only on links belonging to the assigned spanning tree (when variable ϕ_k^s is 1, constraints (5) impose that $x_{ij}^k + x_{ji}^k \leq \beta_{\{ij\}}^s$ and when variable ϕ_k^s is 0, constraints (5) are redundant).

Constraints (7) guarantee that each commodity is assigned to one spanning tree.

Finally, constraints (8) are the capacity constraints that account for the maximum link load on variable μ.

Note that the equality defined by constraints (4) could be used to eliminate variables $\beta_{\{ij\}}^s$ from the model. Nevertheless, they are useful to define the decompositions.

Note also that $|S|$ is un upper limit for the number of STs in the optimal solution (an optimal solution with fewer STs is defined when all variables ϕ_k^s are zero for some values of s).

2.2   Second Compact Model

An alternative MILP model defining the same optimization problem is obtained by disaggregating each variable x_{ij}^k into variables x_{ij}^{ks}, one for each $s \in S$. In this case, variable x_{ij}^{ks} indicates if arc $(i, j) \in A$ is in the path of commodity $k \in K$ in the assigned spanning tree $s \in S$. The resulting MILP model is given by:

$$\min \mu$$

s.t.

$$\sum_{j \in V(i)} \left(y_{ij}^{zs} - y_{ji}^{zs} \right) = \begin{cases} 1, i = z \\ 0, i \neq z \end{cases} \qquad \forall z \in N \backslash \{r\}, \forall s \in S, \forall i \in N \backslash \{r\} \qquad (9)$$

$$\theta_{ij}^s \geq y_{ij}^{zs} \qquad \forall z \in N \backslash \{r\}, \forall (i, j) \in A, \forall s \in S \qquad (10)$$

$$\sum_{j \in V(i)} \theta_{ij}^s = \begin{cases} 1, i \neq r \\ 0, i = r \end{cases} \qquad \forall i \in N, \forall s \in S \qquad (11)$$

$$\theta_{ij}^s + \theta_{ji}^s = \beta_{\{ij\}}^s \qquad \forall \{i, j\} \in A, \forall s \in S \qquad (12)$$

$$x_{ij}^{ks} + x_{ji}^{ks} \leq \beta_{\{ij\}}^s \qquad \forall \{i, j\} \in A, \forall k \in K, \forall s \in S \qquad (13)$$

$$\sum_{j \in V(i)} \left(x_{ij}^{ks} - x_{ji}^{ks} \right) = \begin{cases} \phi_k^s, & i = o_k \\ 0, & i \neq o_k, d_k \\ -\phi_k^s, & i = d_k \end{cases} \forall i \in N, \forall k \in K, \forall s \in S \qquad (14)$$

$$\sum_{s \in S} \phi_k^s = 1 \qquad \forall k \in K \qquad (15)$$

$$\sum_{s \in S} \sum_{k \in K} b_k \left(x_{ij}^{ks} + x_{ji}^{ks} \right) \leq c_{\{ij\}} \mu \qquad \forall \{i, j\} \in A \qquad (16)$$

$$y_{ij}^{zs} \in \{0,1\}, \theta_{ij}^s \in \{0,1\}, \phi_k^s \in \{0,1\}, x_{ij}^{ks} \in \{0,1\}, \beta_{\{ij\}}^s \in \{0,1\} \mu \in [0,1] \ .$$

In this model, the objective function and constraints (9)-(12) and (15) are equal to the first model. Constraints (14) and (16) are straightforward under the proposed variable disaggregation.

Concerning constraints (13), the commodity to spanning tree assignement information, which is defined in the first model by variables ϕ_k^s in constraints (5), is now defined by variables x_{ij}^{ks} and the term $(1 - \phi_k^s)$ of constraints (5) is no longer required on constraints (13). Because of this, the second model is stronger than the first one. In fact, consider the linear relaxation of both models. For example, assume that the spanning tree constraints define two integer spanning trees, $s = 1$ and $s = 2$ and that $\phi_k^1 = \phi_k^2 = 0.5$ for some $k \in K$. Note that in the second model, under these conditions, constraints (13) and (14) guarantee that 50% of the demand for commodity k is routed through tree $s = 1$, while the other 50% is routed through the other tree, $s = 2$. However, in the first model, the right end side of constraints (5) is always at least 0.5, which means that 50% of the demand can be freely routed by the constraint (5) of spanning tree $s = 1$ and the other 50% can be also freely routed by the constraint (5) of spanning tree $s = 2$. This example shows that the convex hull of the first model accommodates solutions which do not belong to that of the second model, yielding the second model a stronger formulation of the first (as will be shown in the computational results).

3   Decompositions

In this section, we describe different decompositions, based on the Dantzig-Wolfe principle [13], in an attempt to obtain more efficient solution techniques, i.e., to achieve the optimal solutions in less nodes generated by the branch-and-bound (B&B) search tree. In this section, we describe the decomposition techniques applied to the second compact model. Their application to the first compact model is analogous with the appropriate adaptations.

The first decomposition considers the set of spanning tree constraints as sub-problems. The second decomposition considers the path constraints as subproblems. The third decomposition is the combination of the two previous ones. Note that the subproblems are either minimum cost spanning tree problems or shortest path problems, for which there are known efficient algorithms. For each decomposition, the model without the subproblem constraints constitutes the master problem and its linear relaxation is solved by column generation, where the columns corresponding to the solutions of the subproblems are dynamically added.

In column generation, the duals of the master problem constraints contribute to the objective function of the subproblems and each solution to a subproblem corresponds to a column of the master problem. This column is inserted in the master problem if the solution yields a negative objective function to the subproblem, in which case we say it is attractive. The master problem is then re-optimized with the new columns and the subproblems are again solved. If no more attractive columns are found, then an optimal solution to the linear relaxation is obtained. At this point, if the solution is not integer, we combine column generation with branch-and-bound (branch-and-price). In each node of the search tree, the branch constraints are added to the master problem and are defined using the original variables. With the new constraints, the master problem is again optimized and the search for attractive columns continues. The branch-and-price process ends when there are no more open nodes of the search tree. For further details on branch-and-price, see [14] and [15].

3.1   Spanning Tree Decomposition

This decomposition considers constraints (9)-(12) as the subproblems (one for each required spanning tree).

Let set P_s be the set of candidate spanning trees for $s \in S$. Attractive solutions are defined by the parameters $\delta_{\{ij\}}^{ps}$, which indicate if edge $\{i,j\} \in A$ is in the candidate spanning tree $p \in P_s$, and have an associated binary variable t_p^s which models if candidate $p \in P_s$ is selected in the optimal solution.

The relationship between $\beta_{\{ij\}}^s$ and t_p^s is

$$\beta_{\{ij\}}^s = \sum_{p \in P_s} \delta_{\{ij\}}^{ps} t_p^s \ .$$

The master problem is given by:

$$\min \mu$$

s.t.

$$\sum_{p \in P_s} t_p^s = 1 \qquad\qquad \forall s \in S \qquad\qquad (17)$$

$$x_{ij}^{ks} + x_{ji}^{ks} \le \sum_{p \in P_s} \delta_{\{ij\}}^{ps} t_p^s \qquad \forall \{i,j\} \in A, \forall k \in K, \forall s \in S \qquad (18)$$

$$\sum_{j \in V(i)} \left(x_{ij}^{ks} - x_{ji}^{ks} \right) = \begin{cases} \phi_k^s, & i = o_k \\ 0, & i \ne o_k, d_k \; \forall i \in N, \forall k \in K, \forall s \in S \\ -\phi_k^s, & i = d_k \end{cases} \qquad (19)$$

$$\sum_{s \in S} \phi_k^s = 1 \qquad\qquad \forall k \in K \qquad\qquad (20)$$

$$\sum_{s \in S} \sum_{k \in K} b_k \left(x_{ij}^{ks} + x_{ji}^{ks} \right) \le c_{\{ij\}} \mu \qquad \forall \{i,j\} \in A \qquad (21)$$

$$t_s^p \in \{0,1\}, \; \phi_k^s \in \{0,1\}, \; x_{ij}^{ks} \in \{0,1\}, \beta_{\{ij\}}^s \in \{0,1\}, \; \mu \in [0,1] \;.$$

Constraints (17) are the added convexity constraints, one for each subproblem, while the other constraints (18), (19), (20), (21) are the ones of the compact model not used as subproblems where variables $\beta_{\{ij\}}^s$ are replaced by the appropriate expression on variables t_p^s.

Consider ν^s to be the dual of the convexity constraint (17) for $s \in S$, and $\pi_{\{ij\}k}^{1s}$, π_{ki}^2, π_k^3, $\pi_{\{ij\}}^4$ to be the duals of the corresponding constraints (18), (19), (20), (21) respectively. The subproblem for each $s \in S$ is given by:

$$\min \sum_{\{i,j\} \in A} \sum_{k \in K} \pi_{\{ij\}k}^{1s} \beta_{\{ij\}}^s - \nu^s$$

s.t.

$$\sum_{j \in V(i)} \left(y_{ij}^{zs} - y_{ji}^{zs} \right) = \begin{cases} 1, i = z \\ 0, i \ne z \end{cases} \forall z \in N \backslash \{r\}, \forall i \in N \backslash \{r\}$$

$$\theta_{ij}^s \ge y_{ij}^{zs} \qquad\qquad \forall z \in N \backslash \{r\}, \forall (i,j) \in A$$

$$\sum_{j \in V(i)} \theta_{ij}^s = \begin{cases} 1, i \ne r \\ 0, i = r \end{cases} \qquad \forall i \in N$$

$$\theta_{ij}^s + \theta_{ji}^s = \beta_{\{ij\}}^s \qquad\qquad \forall \{i,j\} \in A$$

$$y_{ij}^{zs} \in \{0,1\}, \; \theta_{ij}^s \in \{0,1\}, \; \beta_{\{ij\}}^s \in \{0,1\} \;.$$

These subproblems (one for each $s \in S$) define the minimum cost spanning tree problem (link costs given by $\sum_{k \in K} \pi_{\{ij\}k}^{1s}$) which we efficiently solve using the Kruskal algorithm.

3.2 Path Decomposition

This decomposition considers constraints (14) as the subproblems (one for each commodity and for each required spanning tree). Note that the optimal solution of a subproblem can be either a path (with variables x_{ij}^{ks} defining the path and variable ϕ_k^s equal to 1) or a non-path (all variables x_{ij}^{ks} equal to 0 and variable ϕ_k^s also equal to 0).

Let F_{sk} be the set of candidate paths of $k \in K$ to be supported by spanning tree $s \in S$. Let F_{sk}^0 be the set F_{sk} augmented with the non-path. Attractive solutions are defined by the parameters $\eta_{\{ij\}}^{fk}$, which indicate if edge $\{i,j\} \in A$ belongs to path $f \in F_{sk}$, and have an associated binary variable $\varphi_k^{sf} \in \{0,1\}$ which models if candidate $f \in F_{sk}$ is selected in the optimal solution. Non-path solutions are included as initial columns in the master problem and are characterized by all parameters $\eta_{\{ij\}}^{fk}$ being equal to 0.

The relationship between x_{ij}^{ks}, ϕ_k^s and φ_k^{sf} are

$$x_{ij}^{ks} + x_{ji}^{ks} = \sum_{f \in F_{sk}} \eta_{\{ij\}}^{fk} \varphi_k^{sf}$$

$$\phi_k^s = \sum_{f \in F_{sk}} \varphi_k^{sf} \ .$$

The master problem is given by:

$$\min \mu$$

s.t.

$$\sum_{f \in F_{sk}^0} \varphi_k^{sf} = 1 \qquad \forall s \in S, \forall k \in K \tag{22}$$

$$\sum_{j \in V(i)} \left(y_{ij}^{zs} - y_{ji}^{zs} \right) = \begin{cases} 1, i = z \\ 0, i \neq z \end{cases} \qquad \forall z \in N\backslash\{r\}, \forall s \in S, \forall i \in N\backslash\{r\} \tag{23}$$

$$\theta_{ij}^s \geq y_{ij}^{zs} \qquad \forall z \in N\backslash\{r\}, \forall(i,j) \in A, \forall s \in S \tag{24}$$

$$\sum_{j \in V(i)} \theta_{ij}^s = \begin{cases} 1, i \neq r \\ 0, i = r \end{cases} \qquad \forall i \in N, \forall s \in S \tag{25}$$

$$\theta_{ij}^s + \theta_{ji}^s = \beta_{\{ij\}}^s \qquad \forall\{i,j\} \in A, \forall s \in S \tag{26}$$

$$\sum_{f \in F_{sk}} \eta_{\{ij\}}^{fk} \varphi_k^{sf} \leq \beta_{\{ij\}}^s \qquad \forall\{i,j\} \in A, \forall k \in K, \forall s \in S \tag{27}$$

$$\sum_{s \in S} \sum_{f \in F_{sk}} \varphi_k^{sf} = 1 \qquad \forall k \in K \tag{28}$$

$$\sum_{s \in S} \sum_{k \in K} \sum_{f \in F_{sk}} b_k \eta_{\{ij\}}^{fk} \varphi_k^{sf} \leq c_{\{ij\}} \mu \qquad \forall\{i,j\} \in A \tag{29}$$

$$\varphi_k^{sf} \in \{0,1\}, \ y_{ij}^{zs} \in \{0,1\}, \ \theta_{ij}^s \in \{0,1\}, \ \beta_{\{ij\}}^s \in \{0,1\}, \ \mu \in [0,1] \ .$$

Consider ω^{sk} to be the dual of convexity constraint (22) for $k \in K$ and $s \in S$, and π_{zi}^{1s}, $\pi_{z(ij)}^{2s}$, π_i^{3s}, $\pi_{\{ij\}}^{4s}$, $\pi_{k\{ij\}}^{5s}$, π_k^6, $\pi_{\{ij\}}^7$ to be the duals of the corresponding constraints (23), (24), (25), (26), (27), (28), (29) respectively. The subproblem for each $k \in K$ and each $s \in S$ is given by:

$$\min \sum_{\{i,j\}\in A} -\left(\pi^{5s}_{\{ij\}k} + \pi^{7}_{\{ij\}} b_k\right)\left(x^{ks}_{ij} + x^{ks}_{ji}\right) - \pi^{6}_k \phi^s_k - \omega^{sk}$$

s.t.

$$\sum_{j\in V(i)} \left(x^{ks}_{ij} - x^{ks}_{ji}\right) = \begin{cases} \phi^s_k, & i = o_k \\ 0, & i \neq o_k, d_k \ \forall i \in N \\ -\phi^s_k, & i = d_k \end{cases}$$

$$\phi^s_k \in \{0,1\}, \ x^{ks}_{ij} \in \{0,1\} \ .$$

The optimal solution is either when ϕ^s_k is equal to 0 or when ϕ^s_k is equal to 1. In the first case, the objective function value is equal to $-\omega^{sk}$. In the second case, we run a shortest path algorithm (lengths of arcs given by $-\pi^{5s}_{\{ij\}k} - \pi^{7}_{\{ij\}} b_k$) to determine the objective function value. The optimal solution is the best of the two cases.

3.3   Tree and Path Decomposition

This decomposition is the combination of the two previous decompositions. Therefore, all notations previously introduced hold in this case and the master problem is given by:

$$\min \mu$$

s.t.

$$\sum_{p\in P_s} t^s_p = 1 \qquad\qquad \forall s \in S \qquad\qquad (30)$$

$$\sum_{f\in F^0_{sk}} \varphi^{sf}_k = 1 \qquad\qquad \forall s \in S, \forall k \in K \qquad\qquad (31)$$

$$\sum_{f\in F_{sk}} \eta^{fk}_{\{ij\}} \varphi^{sf}_k \leq \sum_{p\in P_s} \delta^{ps}_{\{ij\}} t^s_p \qquad \forall \{i,j\} \in A, \forall k \in K, \forall s \in S \qquad (32)$$

$$\sum_{s\in S} \sum_{f\in F_{sk}} \varphi^{sf}_k = 1 \qquad\qquad \forall k \in K \qquad\qquad (33)$$

$$\sum_{s\in S} \sum_{k\in K} \sum_{f\in F_{sk}} b_k \eta^{fk}_{\{ij\}} \varphi^{sf}_k \leq c_{\{ij\}} \mu \quad \forall \{i,j\} \in A \qquad (34)$$

$$t^s_p \in \{0,1\}, \ \varphi^{sf}_k \in \{0,1\}, \ \mu \in [0,1] \ .$$

Consider ν^s to be the dual of the corresponding constraint (30), ω^{sk} the dual of the corresponding constraint (31), for $s \in S$ and $k \in K$ and $\pi^{1s}_{k\{ij\}}$, π^2_k, $\pi^3_{\{ij\}}$ the duals of the corresponding constraints (32), (33), (34) respectively.

The spanning tree subproblem for each $s \in S$ is given by:

$$\min \sum_{\{i,j\}\in A} \sum_{k\in K} \pi^{1s}_{\{ij\}k} \beta^s_{\{ij\}} - \nu^s$$

s.t.

$$\sum_{j\in V(i)} \left(y^{zs}_{ij} - y^{zs}_{ji}\right) = \begin{cases} 1, i = z \\ 0, i \neq z \end{cases} \forall z \in N\backslash\{r\}, \forall i \in N\backslash\{r\}$$

$$\theta^s_{ij} \geq y^{zs}_{ij} \qquad\qquad \forall z \in N\backslash\{r\}, \forall(i,j) \in A$$

$$\sum_{j \in V(i)} \theta_{ij}^s = \begin{cases} 1, i \neq r \\ 0, i = r \end{cases} \qquad \forall i \in N$$

$$\theta_{ij}^s + \theta_{ji}^s = \beta_{\{ij\}}^s \qquad \forall \{i,j\} \in A$$

$$y_{ij}^{zs} \in \{0,1\}, \ \theta_{ij}^s \in \{0,1\}, \ \beta_{\{ij\}}^s \in \{0,1\}$$

and the path subproblem for each $k \in K$ and each $s \in S$ is given by:

$$\min \sum_{\{i,j\} \in A} - \left(\pi_{\{ij\}k}^{1s} + \pi_{\{ij\}}^3 b_k \right) \left(x_{ij}^{ks} + x_{ji}^{ks} \right) - \pi_k^2 \phi_k^s - \omega^{sk}$$

s.t.

$$\sum_{j \in V(i)} \left(x_{ij}^{ks} - x_{ji}^{ks} \right) = \begin{cases} \phi_k^s, & i = o_k \\ 0, & i \neq o_k, d_k \ \forall i \in N \\ -\phi_k^s, & i = d_k \end{cases}$$

$$\phi_k^s \in \{0,1\}, \ x_{ij}^{ks} \in \{0,1\} \ .$$

Each of these subproblems are solved in the same way as described previously.

4   Computational Results

Computational results for the compact formulations, as well as for the decompositions, were obtained with the networks depicted in Fig. 1 on a Pentium IV machine at 3.4 GHz and with 1GB of RAM. In all cases, we have considered a number of spanning trees $|S| = 2$ and a maximum runtime of 2 hours was allowed.

Two sets of traffic flows were generated for both networks – Set I and Set II for Network A (Table 1) and Set III and Set IV for Network B (Table 2) – where Set I and Set III are significantly more unbalanced (in terms of total traffic per network node) than Set II and Set IV, respectively. All links were considered to have a capacity of 100 Mbps.

In this section, we present in separate subsections the computational results of the compact formulations and the computational results of the decompositions.

4.1   Compact Formulations

The compact formulations for the networks in study were solved using CPLEX 9.1. The results for the first compact model are shown in Table 3, whereas

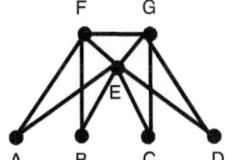
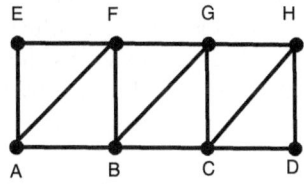

Fig. 1. Networks used as case studies: Network A has 7 nodes and 11 links (left) and Network B has 8 nodes and 13 links (right)

Table 1. Traffic sets generated for Network A

Set I	Set II
A-B 5	A-B 10
A-C 15	A-B 20
A-C 15	C-D 10
B-C 35	C-D 15
B-C 10	A-D 15
B-D 5	A-D 20
B-D 30	B-C 5
B-F 20	B-C 25
C-F 10	B-F 5
C-F 15	C-F 5
D-F 5	D-F 5

Table 2. Traffic sets generated for Network B

Set III	Set IV
A-G 10	A-D 10
A-H 15	A-D 5
A-H 5	A-G 15
B-D 10	A-H 15
B-D 20	A-H 5
B-G 20	B-D 10
B-G 5	B-D 20
B-H 25	B-G 20
E-D 5	B-G 5
E-D 15	E-D 5
E-H 15	E-D 15
E-H 10	E-H 15
G-H 15	E-H 10
G-H 5	G-H 15

the results for the second compact model are shown in Table 4: the results include the LP value (value of the objective function for the Linear Programming relaxation), the MIP value (the value of the objective function of the optimal integer solution), the total computational time (in seconds) and the number of nodes generated by the B&B search tree. The number of B&B search tree nodes is important since most of the resolution time is spent in the B&B process. In the last column, Table 4 also shows the reduction (in percentage) of the number of B&B search tree nodes obtained by the second model.

From the results, we can see that the disaggregation technique used in the second compact model does not improve the LP values. Nevertheless, it reduces a lot the number of B&B nodes (reduction values between 59% and 89%) to prove optimality of the obtained solutions. This has a direct impact on the computational times with similar improvements, yielding the second compact model a more appropriate model to solve this optimization problem.

Table 3. Computational results for the first compact model

	LP	MIP	Time (s)	B&B Nodes
Net A & Set I	0.525	0.55	70.67	39434
Net A & Set II	0.325	0.35	55.05	18998
Net B & Set III	0.467	0.50	3069.78	766878
Net B & Set IV	0.533	0.55	3331.03	564598

Table 4. Computational results for the second compact model

	LP	MIP	Time (s)	B&B Nodes
Net A & Set I	0.525	0.55	51.6	16354 (59%)
Net A & Set II	0.325	0.35	22.6	4992 (74%)
Net B & Set III	0.467	0.50	450.4	81887 (89%)
Net B & Set IV	0.533	0.55	738.7	100952 (82%)

4.2 Decompositions

The proposed decompositions were implemented in C++ using the ADDing (Automatic Dantzig-Wolfe Decomposition for Integer Column Generation) set of C++ classes [16]. The restricted master problems were solved using CPLEX 9.1.

In the implementation of the B&B tree search, we have implemented different node and variable selection strategies.

Concerning the node selection strategy, we have considered three common used search strategies: depth-first, breadth-first and best-bound. On the average of all instances, depth-first and best-bound showed similar performance in the number of generated B&B nodes, while breadth-first had a much worst performance. Since depth-first has lower computational complexity than best-bound, the computational results shown in this paper refer to the results with depth-first strategy.

Concerning variable selection strategy, note first that all integer variables are binary and there are many cases in which when we branch one variable to 1, many other variables become 0. Therefore, we have decided to branch variables first to 1 in order to improve the branch-and-price efficiency. Concerning which variable is to be branched at each time, we have considered a significant number of different strategies both in the order by which the different types of variables are first branched and either taking into account the variables associated to the most loaded links or not.

From the extensive computational tests run with these different variable selection strategies, it was possible to see that the best ones are those where branching of variables $\beta^s_{\{i,j\}}$ are done first. Among these, those that involve branching of variables ϕ^s_k, instead of x^{ks}_{ij} are better, except for the tree decomposition where the performances were similar (note that branching variables $\beta^s_{\{i,j\}}$ and ϕ^s_k are enough to guarantee an integer solution). In the overall, the best strategy was:

(i) closest value to 1 among variables $\beta^s_{\{i,j\}}$ of most loaded links; (ii) if all these variables are integer, then closest value to 1 among the remaining variables $\beta^s_{\{i,j\}}$; (iii) if all variables $\beta^s_{\{i,j\}}$ are integer, then closest value to 1 among variables ϕ^s_k of commodities using most loaded links; (iv) if all these variables are integer, then closest value to 1, among the remaining variables ϕ^s_k. The computational results shown in this paper refer to the results with this variable selection strategy.

Note that CPLEX was run (to solve the compact models) using its automatic settings, since they performed best against any adjustments we attempted to make. Hence, we compare our strategies with the best of CPLEX.

Tables 5 and 6 present the computational results of all three decompositions applied to the first and second compact models, respectively. Each table presents separately the number of B&B nodes and the computational times of all decompositions (in the first case, we repeat the B&B nodes of the CPLEX runs for comparison reasons).

These results show the efficiency of the different decompositions when applied to each compact model.

In the case of the first compact model (Table 5), and for the instances that were solved within the given maximum runtime limit, the tree decomposition has the best performance (both in terms of B&B nodes and in terms of computational time) and the path decomposition has the worst performance. Moreover, the tree decomposition is slightly better than CPLEX in number of B&B nodes (the computational times are not comparable since CPLEX is an highly efficient commercial software package and the branch-and-price approaches were implemented by the authors).

In the case of the second compact model (Table 6), the performance of the tree decomposition, when compared with CPLEX, is roughly the same as in the first model. In this case, though, the path decomposition becomes much more

Table 5. Computational results of decompositions applied to the first compact model

Number of B&B nodes

	CPLEX	Tree	Path	Tree+Path
Net A & Set I	39434	38709	482119	77422
Net A & Set II	18998	14365	184439	49988
Net B & Set III	766878	122186*	207139*	256036*
Net B & Set IV	564598	125117*	174373*	232282*

Computational times in seconds

	Tree	Path	Tree+Path
Net A & Set I	399.64	5224.80	543.80
Net A & Set II	138.08	2323.73	369.30
Net B & Set III	7200	7200	7200
Net B & Set IV	7200	7200	7200

*Reached time limit of 7200 seconds

Table 6. Computational results of the decompositions applied to the second compact model

Number of B&B nodes

	CPLEX	Tree	Path	Tree+Path
Net A & Set I	16354	24461	17571	14222
Net A & Set II	4992	4879	7265	5129
Net B & Set III	81887	79543	70677	65928
Net B & Set IV	100952	35654	27291	39266

Computational times in seconds

	Tree	Path	Tree+Path
Net A & Set I	429.25	150.53	155.67
Net A & Set II	113.17	71.61	75.17
Net B & Set III	4972.30	1247.47	2954.69
Net B & Set IV	3673.50	755.83	2375.39

efficient and, therefore, the best decomposition. Note that although the path decomposition generates on average a little less number of B&B nodes than the tree decomposition, its merit is more apparent in its much shorter computational times. This is because the branch-and-price of path decomposition does not need to introduce too many columns while the branch-and-price of tree decomposition introduces much more columns, penalizing in this way the average computing time per B&B node.

Furthermore, we see that combining tree and path decompositions is not worthy since its performance is, in all cases, between the performance of each individual decomposition.

In general, this kind of exact methods find the optimal solution much sooner than the end of the process (the process ends when the previously best found solution is proven to be optimal). For larger problem instances, that cannot run until the end due to exaggerate computational time, these methods are still useful to find good solutions within some targeted runtime limit. Therefore, it is useful to know when each of the previous methods found the optimal solution.

The proposed decompositions were implemented using a facility provided by ADDing implementation that after solving the linear relaxation by column generation, it solves optimally the master problem with the columns introduced up to that moment and considering the integrality constraints. This solution is an incumbent solution which provides an upper bound to the optimal objective value for the subsequent B&B tree search. Tables 7 and 8 show the B&B node at which each one of the methods found the optimal solution.

These tables show that the path decomposition finds the optimal solution in the very first B&B node, i.e., by solving the master problem with the columns introduced to calculate the LP value. In this decomposition, the subsequent branch-and-price only proves the optimality of such a solution. All the other

Table 7. B&B node of optimal solution - first compact model and associated decompositions

	CPLEX	Tree	Path	Tree+Path
Net A & Set I	402	453	1	27810
Net A & Set II	120	40	1	19701
Net B & Set III	198	3119	1	117374
Net B & Set IV	260	3316	1	165823

Table 8. B&B node of optimal solution - second compact model and associated decompositions

	CPLEX	Tree	Path	Tree+Path
Net A & Set I	477	131	1	119
Net A & Set II	460	100	1	129
Net B & Set III	70	67	1	45
Net B & Set IV	40	858	1	2026

methods (solving the compact models with CPLEX and the other decompositions) take some B&B nodes to find the optimal solution.

5   Conclusion

In this paper, we have addressed the minimization of the maximum link load of telecommunication networks based on multiple spanning tree routing. Two mixed integer linear programming models defining the optimization problem were proposed and compared. The second model, which is derived from the first one based on a variable disaggregation technique, was found to be much more efficient with average computational improvements of 76%. Three different decompositions, based on the Dantzig-Wolfe decomposition principle, that can be applied to both models were also proposed and compared. When applied to the first model, the tree decomposition outperforms the others but when applied to the more efficient second model, the path decomposition is the best technique amongst all. Moreover, the path decomposition applied to the second model is also the technique that finds the optimal incumbent in much shorter computational times, making it also the best option to run as an heuristic for larger problem instances that cannot be solved to optimality.

Acknowledgment

The authors would like to thank the portuguese FCT (Fundação para a Ciência e a Tecnologia) for its support through projects PTDC/EIA/64772/2006 and POSC/EIA/57203/2004 and through the post-doc grant SFRH/BPD/41581/ 2007 of the first author.

References

1. IEEE Standard 802.1s, Virtual Bridged Local Area Networks - Amendment 3: Multiple Spanning Trees (2002)
2. MEF Technical Specification 6.1, Ethernet Services Definitions -Phase 2. Metro Ethernet Forum (2008)
3. de Sousa, A.F., Soares, G.: Improving Load Balance and Minimizing Service Disruption on Ethernet Networks using IEEE 802.1S MSTP. In: EuroFGI Workshop on IP QoS and Traffic Control, pp. 25–35. IST Press (2007)
4. Padmaraj, M., Nair, S., Marchetti, M., Chiruvolu, G., Ali, M.: Traffic Engineering in Enterprise Ethernet with Multiple Spanning Tree Regions. In: Proc. of System Communications (ICW 2005), Montreal, Canada, pp. 261–266 (2005)
5. de Sousa, A.F., Soares, G.: Improving Load Balance of Ethernet Carrier Networks using IEEE 802.1S MSTP with Multiple Regions. In: Boavida, F., Plagemann, T., Stiller, B., Westphal, C., Monteiro, E. (eds.) NETWORKING 2006. LNCS, vol. 3976, pp. 1252–1260. Springer, Heidelberg (2006)
6. Kern, A., Moldovan, I., Cinkler, T.: Scalable Tree Optimization for QoS Ethernet. In: IEEE Symp. on Computers and Communications (ISCC 2006), pp. 578–584 (2006)
7. Ali, M., Chiruvolu, G., Ge, A.: Traffic Engineering in Metro Ethernet. IEEE Network 19(2), 10–17 (2005)
8. Kolarov, A., Sengupta, B., Iwata, A.: Design of Multiple Reverse Spanning Trees in Next Generation of Ethernet-VPNs. In: IEEE GLOBECOM 2004, vol. 3, pp. 1390–1395 (2004)
9. Sharma, S., Gopalan, K., Nanda, S., Chiueh, T.: Viking: A Multi-Spanning-Tree Ethernet Architecture for Metropolitan Area and Cluster Networks. In: IEEE INFOCOM 2004, vol. 4, pp. 2283–2294 (2004)
10. Ishizu, K., Kuroda, M., Kamura, K.: SSTP: an 802.1s Extention to Support Scalable Spanning Tree for Mobile Metropolitan Area Network. In: IEEE GLOBECOM 2004, vol. 3, pp. 1500–1504 (2004)
11. Lim, Y., Yu, H., Das, S., Lee, S.-S., Gerla, M.: QoS-aware multiple spanning tree mechanism over a bridged LAN environment. In: IEEE GLOBECOM 2003, vol. 6, pp. 3068–3072 (2003)
12. de Sousa, A.F.: Improving Load Balance and Resilience of Ethernet Carrier Networks with IEEE 802.1S Multiple Spanning Tree Protocol. In: Int. Conference on Networking (ICN), pp. 95–102. IEEE Xplore (2006)
13. Dantzig, G.B., Wolfe, P.: Decomposition principle for linear programs. Operations Research 8, 101–111 (1960)
14. Barnhart, C., Johnson, E.L., Nemhauser, G.L., Savelsbergh, M.W.P., Vance, P.H.: Branch-and-price: Column generation for solving huge integer programs. Operations Research 46, 316–329 (1998)
15. Vanderbeck, F.: On Dantzig-Wolfe decomposition in integer programming and ways to perform branching in a branch-and-price algorithm. Operations Research 48(1), 111–128 (2000)
16. Alvelos, F.: Branch-and-price and multicommodity flows. PhD Thesis, Universidade do Minho (2005)

Local Restoration for Trees and Arborescences

Paola Iovanna[1], Gaia Nicosia[2], Gianpaolo Oriolo[3],
Laura Sanità[3], and Ezio Sperduto[2]

[1] Ericsson Lab Italy
Via Anagnina 203, 00040, Roma, Italy
paola.iovanna@ericsson.com
[2] Dipartimento di Informatica e Automazione
Università degli studi Roma Tre
{nicosia,sperduto}@dia.uniroma3.it
[3] Dipartimento di Ingegneria dell'Impresa
Università degli studi di Roma Tor Vergata
{oriolo,sanita}@disp.uniroma2.it

Abstract. Protocols belonging to the Spanning Tree Protocol (STP)
route traffic demands on tree topologies that are evaluated through shortest path procedures. In this paper we deal with the problem of assigning
costs to the arcs of a network in order to guarantee that SPT protocols
efficiently re-route traffic demands in failure situations: namely, without
redirecting traffic demands that are not affected by the failure. We say
that a communication network has the *local tree-restoration* property if
there exists a set of costs for its arcs such that the above property holds.

We show that an undirected network has the local tree-restoration
property if and only if it is 2-connected. In particular, we provide a quite
simple procedure for assigning costs to the arcs of a 2-connected network
so that the property holds. For the directed case, we show that deciding
whether a network has the local tree-restoration property is NP-hard,
even in some "simple" cases.

1 Introduction

A crucial issue in the design of new generation communication networks is the
survivability, that is the capability of a network to remain operational even
when a single component, like a link, fails. Therefore, *fault-tolerant* communication networks are required, to guarantee that traffic demands can be efficiently
re-routed even in failure situations. In particular, since re-routing of traffic can
often be expensive and/or may cause long delay in the transmission [2,5], in
order to meet QoS constraints, a key property of the re-routing strategy is the
so-called *strong resilience* [1] or *minimum service disruption property* [8], requiring that traffic, which is not affected by the failure, is not redirected in the
process of restoring the affected traffic. In other words, under the minimum service disruption property, the number of traffic demands that are re-routed is
minimized.

R. Valadas and P. Salvador (Eds.): FITraMEn 2008, LNCS 5464, pp. 130–140, 2009.
© Springer-Verlag Berlin Heidelberg 2009

In this paper we deal with the problem of designing tree-based routing strategies that are strongly resilient. Our interest is sparkled by routing protocols belonging to the Spanning Tree Protocol (STP)[11] family, used for instance in Ethernet networks, where tree topologies are used to route traffic demands and shortest path procedures are used to evaluate these trees. An STP routing protocol routes the traffic on a *spanning tree*, namely, the shortest path tree for some special node r of the network, called the *root*, with respect to suitable arc costs set by network operators.

More formally, we consider the following setting. We are given a communication network $G(V, E)$, that we first suppose to be undirected. The links of E may fail; in particular, we consider the case of a single link failure (in fact, assuming that a link that is down recovers quickly, the probability of having several link failures at the same time is negligible). There are, therefore, $|E| + 1$ *scenarios*, each corresponding either to the failure of some link $e \in E$, the *back-up scenario*, or to the no-failures case, the *primary scenario*.

For each scenario, traffic will be routed on a *spanning tree* of the network that is operational in that scenario. Therefore, in the primary scenario, traffic will be routed on a spanning tree $T(\emptyset)$ of G; in the back-up scenario corresponding to the failure of link $e \in E$, traffic will be routed on a spanning tree $T(e)$ of $G(V, E\setminus\{e\})$. (In the following, we indicate with $G \setminus \{e\}$ the subgraph $G(V, E \setminus \{e\})$.)

Note that, for each scenario, we *cannot* directly choose the spanning tree, that has to be a *shortest path tree*, rooted at a node r, for the network that is operational in that scenario, with respect to arc costs $w : E \mapsto \mathcal{Z}_+$. Therefore, $T(\emptyset)$ has to be a shortest path tree of G, while $T(e)$, $e \in E$, has to be a shortest path of $G(V, E \setminus \{e\})$. On the other hand, we *can* choose the *root* r and the *cost vector* w; note that both r and w have to be the same for each scenario.

The spanning tree $T(\emptyset)$ is called the *primary tree*; the paths in the primary tree are called *primary paths* and the primary path between u and v is denoted by $P_{uv}(\emptyset)$. Note that requiring the minimum service disruption property amounts to requiring that, for each arc e in the primary tree, the following holds:

for each pair of nodes u, v that do not use e in their primary path, the primary path is still operational when e fails, i.e. $P_{uv}(\emptyset) \subseteq T(e)$, if $e \notin P_{uv}(\emptyset)$.

In the following, if the primary tree and the back-up trees determined by a pair (r, w) satisfy the above property, we say that the pair (r, w) has the *local tree-restoration property*. Analogously, a network G has the local tree-restoration property if there exists a pair (r, w) with the property. Finally, we say that a spanning tree $T \subseteq G$ has the *local tree-restoration property*, if there exists a pair (r, w) with the property, such that T is the primary tree defined by (r, w).

Our definitions directly extend to the case in which G is directed. In this case, we assume that each traffic demand is directed from a root r to some node v and therefore, instead of spanning trees, we deal with spanning *arborescences*, i.e. directed spanning trees, rooted at r (note that r is now given from the outset and cannot be chosen).

1.1   Related Works

In the past years, fault-tolerant networks have been studied intensively (see [3] for a survey). The problem of finding the best link which re-connects the communication topology (*swap arc*) with efficient techniques was studied by Nardelli *et al.* [5] and, with distributed algorithms, by Flocchini *et al.* [2].

In the context of STP family protocols, in the last years efficient protocols have been standardized, such as Rapid Spanning Tree Protocol (RSTP)[13] and Multiple Spanning Tree Protocol (MSTP)[12], therefore the design of networks based on such routing protocols is receiving a lot of attention (see e.g. [4,6,7]).

In particular, the problem of minimum service disruption in networks using the Multiple Spanning Tree Protocol has been addressed by De Sousa *et al.* in [8]. They provide an algorithm that, given any spanning tree T of a 2-arc-connected undirected network and a root r, computes arcs costs w such that the pair (r, w) has the local tree-restoration property and T is the primary tree defined by (r, w). Therefore, according to our definitions, the following result is proved in [8]: every spanning tree of a 2-arc-connected undirected network G has the local tree-restoration property.

1.2   Our Results

We first deal with undirected networks. We sharpen the result in [8] and show that an undirected network G has the local tree-restoration property if and *only if* it is 2-arc-connected. We also give a different proof of the fact that *every* spanning tree T of a 2-arc-connected undirected network G has the local tree-restoration property: the proof includes a very simple procedure for assigning the costs on the arcs so that T has the local tree-restoration property.

We then deal with directed networks. We show that deciding whether a directed network G has the local tree-restoration property is NP-hard (therefore, this decision problem can unlikely be solved by an algorithm with time-complexity bounded by a polynomial in the size of the network), already when each node of G has distance at most 2 from the root r. On the sunny side, the problem is easy when each node has distance 1 from r and it is easy to check whether a *given* spanning arborescence T has the local tree-restoration property.

The paper is organized as follows. We close this section with a few definitions. In the next section we deal with undirected networks and show that G has the local tree-restoration property if and only if it is 2-arc-connected and provide a simple procedure for assigning costs on the arcs of the network for obtaining, through the SPT protocol, spanning trees with the local tree-restoration property. In Section 3, we show that the problem of deciding whether a directed network G has the local tree-restoration property is NP-hard. Finally, in Section 4 future research directions are discussed and conclusions are drawn.

An undirected network G is *2-(arc)-connected* if, for each pair of nodes, there are two arc disjoint paths connecting them (for the sake of simplicity, in the following, we will use the term 2-connected instead 2-(arc)-connected). For a subset of nodes $S \subseteq V$ of an undirected network $G(V, E)$, we denote by $\delta_G(S)$

the set of arcs in $E(G)$ with exactly one endpoint in S. For a subset of nodes $S \subseteq N$ of a directed network $G(N, A)$, we denote by $\delta_G^+(S)$ the set of arcs in $A(G)$ outgoing S, and $\delta_G^-(S)$ the set of arcs incoming S.

2   Local Tree-Restoration for Undirected Networks

In this section we deal with undirected networks. First of all we provide a reformulation of the local tree-restoration property that does not involve the definition of the root and the arcs costs.

Lemma 1. *Let* $G(V, E)$ *be an undirected network. A spanning tree* T *has the local tree-restoration property if and only if, for each* $e \in E(T)$, *there exists* $f \in E(G) \setminus E(T)$ *such that* $T(e) = (T \setminus \{e\}) \cup \{f\}$ *is a spanning tree.*

Proof. Necessity is trivial.

Sufficiency. Choose r arbitrarily. Number arbitrarily all arcs $f \in E(G) \setminus E(T)$ from 1 to $k = |E(G) \setminus E(T)|$ and let i_f be the index, or number, associated to arc f. Consider the following cost function on the arcs:

$$w(e) = \begin{cases} 1 & \text{if } e \in E(T) \\ n \cdot i_e & \text{if } e \in E(G) \setminus E(T) \end{cases}$$

We first show that the following statements hold, with respect to the cost vector w: (i) T is the unique shortest path tree rooted at r of G; (ii) for each $e \in E(T)$, there exists $f \in E(G) \setminus E(T)$ such that $(T \setminus \{e\}) \cup \{f\}$ is the unique shortest path tree rooted at r of the network $G \setminus \{e\}$.

Proof of (i). Assume that T is not the unique shortest path tree rooted at r of G. Then, there must exist a path from r to some node v that is no longer than the $r - v$ path in T and uses at least one arc f not belonging to $E(T)$. This yields to a contradiction, since such path has a cost greater than $n - 1$, while (r, v)-path in T has cost less or equal than $n - 1$.

Proof of (ii). Let e be an arc of T and let S_e and $\bar{S}_e$ be the nodes in the two components of $T \setminus \{e\}$, with $r \in S_e$. Let f be the minimum cost arc in $\delta_{G \setminus \{e\}}(S_e)$. We claim that $T(e) = (T \setminus \{e\}) \cup \{f\}$ is the unique shortest path tree rooted at r of $G \setminus \{e\}$. Suppose the contrary. Then, in $G \setminus \{e\}$ there must exist a path P from r to some node v that is no longer than the (r, v)-path in $T(e)$ and uses at least one arc h not belonging to $E(T(e))$. If $v \in S_e$, P has a cost greater than $n - 1$ while the (r, v)-path in $T(e)$ has cost less than or equal to $n - 1$, a contradiction. Viceversa, if $v \in \bar{S}_e$, P must use at least one arc from $\delta_{G \setminus \{e\}}(S_e)$ and its cost is greater than $(i_f + 1)n - 1$: this is trivial if P uses some arc from $\delta_{G \setminus \{e\}}(S_e)$ different from f, else it follows from the fact that P uses f and an arc in $E(G) \setminus E(T)$. Since the (r, v)-path in $T(e)$ has cost less or equal than $i_f n + n - 1$, we have again a contradiction.

Finally, it is easy to see that (i) and (ii) imply that the pair (r, w) has the local tree-restoration property, and since T is the primary tree defined by (r, w), also T has the local tree-restoration property.

The previous lemma reduces the problem of checking whether a spanning tree T of a network G has the local tree-restoration property to checking whether the following property holds:

($\star$) For each arc $e \in E(T)$, there exists $f \in E(G) \setminus E(T)$ such that $T(e) = (T \setminus \{e\}) \cup \{f\}$ is a spanning tree.

An example of an undirected network (and a spanning tree) with the local tree-restoration property is depicted in Figure 1.

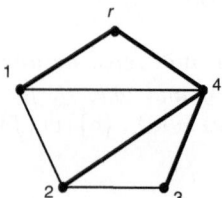

Fig. 1. An undirected network with the local tree-restoration property

The next lemma shows that the above property holds indeed for *every* spanning tree T, when G is 2-connected. Also 2-connection of G is a necessary condition for the above property to hold. We recall that checking whether a network is 2-connected can be easily performed in linear time with a Depth-First-Search of the network [10].

Lemma 2. *Let $G(V, E)$ be an undirected network. If G is 2-connected then ($\star$) holds for each spanning tree T. Vice versa, if ($\star$) holds for some spanning tree T of G, then G is 2-connected.*

Proof. First suppose that G is 2-connected and let T be a spanning tree of G. Let e be an arc of T and $(S_e, \bar{S}_e)$ the subsets of nodes in the two components of $T \setminus \{e\}$. Since G is 2-connected, there exists $f \in \delta_G(S_e)$, $f \neq e$. Then, $T(e) = (T \setminus \{e\}) \cup \{f\}$ is a spanning tree and we are done.

Now suppose that ($\star$) holds for some spanning tree T of G. If G is not 2-connected, then there exists an arc $e \in E$ such that $G \setminus \{e\}$ is not connected. Trivially, ($\star$) cannot hold with respect to e, a contradiction.

If we put together Lemma 1 and Lemma 2 we get the main result of this section:

Theorem 1. *An undirected network has the local tree-restoration property if and only if is 2-connected. Moreover, each spanning tree of a 2-connected network has the local tree-restoration property.*

We want to emphasize that the proof of Lemma 1 shows a simple procedure for endowing, in 2-connected networks and through the SPT protocol, a spanning tree T with the local tree-restoration property. We recap this procedure in the following.

First, we arbitrarily choose a root r. Then we arbitrarily number all arcs $f \in E(G) \setminus E(T)$ from 1 to $k = |E(G) \setminus E(T)|$ and let i_f be the number associated to arc f. Finally, we define the following cost function on the arcs:

$$w(e) = \begin{cases} 1 & \text{if } e \in E(T) \\ n \cdot i_e & \text{if } e \in E(G) \setminus E(T) \end{cases}$$

3   Local Tree-Restoration for Directed Networks

In this section we extend the local tree-restoration property to the scenario in which the communication network is a directed network $G(N, A)$. To avoid confusion, we recall here the definitions stated in the introduction for directed networks. Let r be a node in $N(G)$ and w be a cost function $w : A \rightarrow Z_+$.

We say that the pair (r, w) has the local tree-restoration property, if the following hold: (i) there exists a unique shortest path arborescence T (primary arborescence) rooted at r, (ii) for each $a \in A(T)$, in the network $G(N, A \setminus \{a\})$ there exists a unique shortest path arborescence $T(a)$, rooted at r, such that for all the directed paths $P_{uv} \subseteq T$ from u to v that do not contain the arc a, we have that $P_{uv} \subseteq T(a)$.

Similarly, a spanning arborescence $T \subseteq G$ has the local tree-restoration property, if there exists a pair (r, w) with the property, such that T is the primary arborescence defined by (r, w).

Finally, a pair (G, r), with $r \in N(G)$, has the local tree-restoration property if there exists a spanning arborescence T, rooted at r, that has the local tree-restoration property.

As in the previous section, we start with a reformulation of the local tree-restoration property that does not involve the definition of the arc costs.

Lemma 3. *Let $G(N, A)$ be a directed network and $r \in N$. A spanning arborescence T, rooted at r, has the local tree-restoration property if and only if, for each $a \in A(T)$, there exists $f \in A(G) \setminus A(T)$ such that $T(a) = (T \setminus \{a\}) \cup \{f\}$ is a spanning arborescence rooted at r.*

Proof. Necessity is trivial.

Sufficiency. Number the arcs in $A(G) \setminus A(T)$ from 1 to $|A(G) \setminus A(T)|$ so that the following property holds: for each sub-arborescence T_v of T with root $v \neq r$, the number, or index, i_a associated to every arc $a \in \delta_G^-(v)$ is smaller than the index i_f associated to every arc $f \in \delta_G^-(u)$, if u is a descendant of v in T_v. Notice that this can be easily done in polynomial time, e.g. considering first all nodes v having distance 1 from r in the tree T and numbering the arcs $a \in \delta_G^-(v)$, then all the nodes v having distance 2 from r, and so on. With this numbering, we apply the same cost function w of Lemma 1.

Let us prove that w is such that (i) T is the unique shortest path arborescence with root r for network G and (ii) for each $a \in A(T)$, there exists $f \in A(G) \setminus A(T)$ such that $T(a) = (T \setminus \{a\}) \cup \{f\}$ is the unique shortest path arborescence rooted at r for the network $G \setminus \{a\}$.

Proof of (i). It is a simple extension of the proof of Lemma 1.

Proof of (ii). Let a be an arc of T and let S_a and $\bar{S}_a$, with $r \in S_a$, be the subsets of nodes respectively in the first and second component of $T \setminus \{a\}$. Let $v \in \bar{S}_a$ be the root of the sub-arborescence T_v that is the second component of $T \setminus \{a\}$.

We know that there exists an arc f such that $T(a) = (T \setminus \{a\}) \cup \{f\}$ is an arborescence: necessarily, $f \in \delta_{G \setminus \{a\}}(v)$, since an arborescence must have one arc incoming in v. Let f be the minimum cost arc in $\delta_{G \setminus \{a\}}(v)$. Notice that f is also the minimum cost arc in $\delta^-_{G \setminus \{a\}}(\bar{S}_a)$: in fact, an arc in $\delta^-_{G \setminus \{a\}}(\bar{S}_a)$ either ends in v or it ends in a descendant of v, but in the latter case, the cost function w implies that this arc has cost greater than $w(f)$. With this observation, proving that $T(a)$ is the unique shortest path arborescence in $G \setminus \{a\}$ rooted at r is then an extension of the proof of Lemma 1.

An example of a directed network (and a spanning arborescence) with the local tree-restoration property is illustrated in Figure 2.

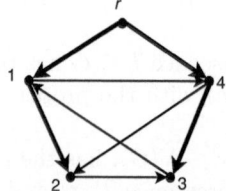

Fig. 2. A directed network with the local tree-restoration property

Building upon the previous lemma, we define another reformulation of the local tree-restoration property for some spanning arborescence T of a directed network G.

Lemma 4. *Let $G(N, A)$ be a directed network and T a spanning arborescence rooted at a node r of G. T has the local tree-restoration property if and only if, for each $v \in N(T)$, there exists $(u, v) \in A(G) \setminus A(T)$ such that u is not a descendant of v in T.*

Proof. We prove that the above statement is equivalent to requiring that, for each $a \in A(T)$, there exists $f \in A(G) \setminus A(T)$ such that $T(a) = (T \setminus \{a\}) \cup \{f\}$ is a spanning arborescence rooted at r. Then one can use lemma 3 to conclude.

First we prove necessity. Let $a = (u, v) \in A(T)$ and $S_a, \bar{S}_a$ be the sets of nodes in the two components associated to the removal of a from T, with $r \in S_a$. If in $G \setminus \{a\}$ all the arcs incoming in v come from a descendant of v in T, then there exists no arc $(u, v) \in A(G) \setminus A(T)$, with $u \in S_a$. Therefore, in any spanning arborescence of $G \setminus \{a\}$, rooted in r, there are at least two arcs that do not belong to T, namely, one connecting S_a and $\bar{S}_a$, and the other entering v. So, in this case, there is no arc $f : (T \setminus \{a\}) \cup \{f\}$ is an arborescence rooted at r.

Now we prove sufficiency. Let $f = (u, v)$ be an arc such that u is not a descendant of v in T. Note that $u \in S_a$, therefore (u, v) reconnects the components S_a and $\bar{S}_a$. It follows that $T(a) = T \setminus \{a\} \cup \{f\}$ is a spanning arborescence. The result follows.

We point out that both the previous lemmas provide an easy way to check in polynomial time if a given arborescence T of a directed network G has the local tree-restoration property. Also, when T has the local tree-restoration property, the proof of Lemma 3 also provides a simple procedure for assigning costs to the arcs of G as to define T through the SPT protocol.

3.1 Existence of Arborescences with Local Tree-Restoration Property

For a directed network $G(N, A)$ the *distance* of a node $v \in V$ from r is the minimum number of arcs (*hops*) in a directed path from r to v. We denote by $k(G, r)$ the *maximum distance* of a node $v \in V$ from r.

We show in the following that, deciding whether a pair (G, r) has the local tree-restoration property can be done in polynomial time if $k(G, r) = 1$, while it is NP-hard otherwise, even for $k(G, r) = 2$.

The following lemma allows us to restrict to some particular arborescences.

Lemma 5. *A pair (G, r) has the local tree-restoration property if and only if there exists an arborescence T with the local tree-restoration property such that, for each node u at distance one from r, the arc $(r, u) \in A(T)$.*

Proof. Sufficiency is trivial.

Necessity. We show how to construct such an arborescence T, given any arborescence $\bar{T}$ that has the local tree-restoration property. Suppose that there is a node $v : d_G(r, v) = 1 \neq d_{\bar{T}}(r, u)$ (otherwise we are done). Let $f = (r, v) \notin A(\bar{T})$, and let $a \in A(\bar{T})$ be the arc incoming in v in the tree T. Finally, let $T := (\bar{T} \setminus \{a\}) \cup \{f\}$. It is easy to see that T is an arborescence rooted at r. We claim that T has the local tree-restoration property. Then, we one can use induction to conclude.

Using Lemma 4, it is enough to check if every node $v \neq r$ has an incoming arc (u, v) for some u that is not a descendant of v in T. Notice that, for each node $u \in N$, the set of nodes $D_T(u)$ that are descendant of u in T is *contained* in the set of nodes $D_{\bar{T}}(u)$ that are descendant of u in $\bar{T}$ (i.e. $D_T(u) \subseteq D_{\bar{T}}(u)$). Therefore, since $\bar{T}$ satisfies the hypothesis of Lemma 4, it follows that this holds also for T. The result follows.

We are now ready to state our first result.

Theorem 2. *Let $G(N, A)$ be a directed network and $r \in N$ such that $k(G, r) = 1$. Deciding if (G, r) has the local tree-restoration property can be done in polynomial time.*

Proof. Let T be the arborescence rooted at r, such that $A(T) := \{(r,v), \text{ for all } u \in N \setminus \{r\}\}$. From Lemma 5, we know that if (G,r) has the local tree-restoration property, then T has the local tree-restoration property. We check in polynomial time if T has the local tree-restoration property using Lemma 4. The result follows.

We close the paper with the following hardness result.

Theorem 3. *Let $G(N,A)$ be a directed network and $r \in N$. Deciding if (G,r) has the local tree-restoration property is NP-hard.*

Proof. In Lemma 4 we have shown that a spanning arborescence T has the local tree-restoration property if and only if for each $v \in N \setminus \{r\}$ there is an arc $(u,v) \in A(G) \setminus A(T)$ such that u is not a descendant of v. We will prove here that, given a network $G(N,A)$ and a node $r \in N$, deciding if there exists a spanning arborescence T rooted at r with the latter property is NP-complete, by using a reduction from k-coloring problem. Let us restate both problems as decision problems.

> **Problem LTR:** Given a directed graph $G' = (N,A)$ and a root node $r \in N$, is there an arborescence T such that the following holds: for each $v \in N \setminus \{r\}$ there is an arc $(u,v) \in A(G') \setminus A(T)$ such that u is not a descendant of v in T?

> **Problem k-COL:** Given an undirected graph $G = (V,E)$ and an integer $k > 0$, does there exist a function $c : V \mapsto \{1,2,\ldots k\}$ such that $c(u) \neq c(v)$ for each $\{u,v\} \in E$?

Given an instance (G,k) of k-COL problem, let $|V(G)| = n$ be the number of nodes and $|E(G)| = m$ the number of arcs. We build the corresponding instance (G',r) of LTR problem as follows. The set of nodes $N(G')$ includes, besides all the nodes $v_h \in V(G)$, a root node r, a node c_j^0, for each color $j = 1,\ldots,k$, and a node c_j^i for each arc e_i, $i = 1,\ldots,m$, and each color j, $j = 1,\ldots,k$. Formally,

$$N(G') = \{r\} \cup \{v_1,\ldots,v_n\} \cup$$
$$\{c_j^i : j = 1,2,\ldots k \text{ and } i = 0,1,\ldots m\}.$$

The set of arcs $A(G')$ is defined as follows:

$$A(G') = \{(r,c_j^0) : j = 1,2,\ldots,k\} \cup$$
$$\{(c_j^0, c_{j+1}^0) : j = 1,\ldots,k-1\} \cup \{(c_k^0, c_1^0)\} \cup$$
$$\{(c_j^i, c_j^{i+1}) : i = 0,1,\ldots,m-1; j = 1,\ldots k\} \cup$$
$$\{(c_j^m, v_h) : j = 1,\ldots,k; h = 1,\ldots,n\} \cup$$
$$\{(v_h, c_j^i) : v_h \text{ is an endpoint of } e_i \in E(G),$$
$$h = 1,\ldots,n; \ j = 1,2,\ldots,k\}.$$

Suppose $k \geq 2$ (otherwise k-COL problem is trivial). We show that a solution for the instance (G, k) of k-COL implies the existence of a solution for the instance (G', r) of LTR and vice versa.

Given a solution of k-COL, we build a spanning arborescence T rooted at r with the local tree-restoration property in the following way. T will include (i) arcs (r, c_j^0) for $j = 1, \ldots, k$; (ii) arcs (c_j^i, c_j^{i+1}) for $i = 0, \ldots, m-1$ and $j = 1, \ldots, k$; (iii) an arc (c_j^m, v_h) for each node v_h, $h = 1, \ldots, n$, if and only if v_h is colored with color j in the solution of k-COL on G, i.e. if and only if $c(v_h) = j$. Trivially, T is a spanning arborescence on G' rooted at r.

Now we show that for each node $v \in N(G') \setminus \{r\}$ there exist an arc $f = (u, v)$ such that u is not a descendant of v in T. For a node c_j^0, such an arc is $f = (c_{j-1}^0, c_j^0)$ if $j \neq 1$, else, $f = (c_k^0, c_1^0)$. For a node v_h, such an arc can be (c_j^m, v_h) for some $j \neq c(v_h)$. Finally, for a node c_j^i, with $i \geq 1$, we can choose f as one between the two arcs $(v_{\bar{l}}, c_j^i)$ and $(v_{\bar{h}}, c_j^i)$ incoming c_j^i, since by construction at least one among $v_{\bar{l}}$ and $v_{\bar{h}}$ is not a descendant of c_j^i in T: in fact, recall that $e_i = \{v_{\bar{h}}, v_{\bar{l}}\} \in E(G)$ and at most one can have the color j in the solution of k-COL.

Now we show that a solution for the instance (G', r) of LTR implies the existence of a solution for the instance (G, k) of k-COL. Suppose we have an arborescence $T \subseteq G'$ rooted at r that is a solution of LTR. It follows from Lemma 5 that, without loss of generality, we can assume that $(r, c_j^0) \in A(T)$ for $j = 1, \ldots, k$. Let P_{rv_h} be the path from node r to a node v_h defined by T. Observe that this path must contain all the arcs $(c_j^0, c_j^1), \ldots (c_j^{m-1}, c_j^m)$ for exactly one j in $1, .., k$ (in general, j depends on v_h). We claim that assigning the color j to the node v_h, for each $h = 1, \ldots, n$, leads to a feasible solution for the instance (G, k) of the k-COL problem. Suppose the contrary. Then there must be at least a pair of nodes (say v_h and v_l) that have been assigned to a same color j and that are connected by an arc $e_i = \{v_h, v_l\} \in E(G)$. This means that both nodes v_h and v_l are descendant of c_j^i in T. Since the only arc incoming node c_j^i, different from (v_h, c_j^i) and (v_l, c_j^i), is $(c_j^{i-1}, c_j^i) \in A(T)$, it follows that there is not any arc $(u, c_j^i) \in A(G') \setminus A(T)$ such that u is not a descendant of c_j^i in T, i.e. T is not a solution for the instance (G', r), a contradiction.

The above result can be strengthened by proving that problem LTR remains NP-hard even when $k(G, r) = 2$.

Theorem 4. *Let $G(N, A)$ be a directed network and $r \in N$ such that $k(G, r) \geq 2$. Deciding if (G, r) has the local tree-restoration property is NP-hard.*

The proof of this case is rather technical and we refer to [9] for details.

4   Conclusions

In this paper we have addressed the problem of verifying whether a directed or undirected network has the local tree-restoration property. We have proved that for undirected networks this can be done in polynomial time, while for directed networks the problem is NP-hard, even on networks with a very special structure.

A possible topic for further research is to check whether our positive results for the undirected networks extend to the case where more links fail at the same time.

References

1. Brightwell, G., Oriolo, G., Sheperd, F.B.: Reserving Resilient Capacity in a Network. Siam Journal on Discrete Mathematics 14(4), 524–539 (2001)
2. Flocchini, P., Pagli, L., Prencipe, G., Santoro, N., Widmayer, P.: Computing all the best swap edges distributively. Journal of Parallel and Distributed Computing 68(7) (2008)
3. Grötschel, M., Monma, C.L., Stoer, M.: Design of survivable networks. In: Handbook in OR and MS, vol. 7, pp. 617–672. Elsevier, Amsterdam (1995)
4. Kolarov, A., Sengupta, B., Iwata, A.: Design of Multiple Reverse Spanning Trees in Next Generation of Ethernet-VPNs. In: IEEE GLOBECOM 2004, vol. 3, pp. 1390–1395 (2004)
5. Nardelli, E., Proietti, G., Widmayer, P.: Swapping a Failing Edge of a Single Source Shortest Paths Tree Is Good and Fast. Algorithmica 35, 56–74 (2003)
6. Padmaraj, M., Nair, S., Marchetti, M., Chiruvolu, G., Ali, M.: Traffic Engineering in Enterprise Ethernet with Multiple Spanning Tree Regions. In: Proc. of System Communications (ICW 2005), Montreal, Canada, pp. 261–266 (2005)
7. Sharma, S., Gopalan, K., Nanda, S., Chiueh, T.: Viking: A Multi-Spanning-Tree Ethernet Architecture for Metropolitan Area and Cluster Networks. In: IEEE INFOCOM 2004, vol. 4, pp. 2283–2294 (2004)
8. de Sousa, A.F., Soares, G.: Improving Load Balance and Minimizing Service Disruption on Ethernet Networks using IEEE 802.1S MSTP. In: Proc. EuroFGI Workshop on IP QoS and Traffic Control, Lisbon, Portugal, vol. 1, pp. 25–35 (2007)
9. Sperduto, E.: Combinatorial structures in communication networks, Ph. D. thesis in Computer Science and Automation, Università Roma Tre (2008)
10. Tarjan, R.: Depth-first search and linear graph algorithms. SIAM Journal on Computing 1(2), 146–160 (1972)
11. Standard IEEE 802.1D
12. Standard IEEE 802.1s
13. Standard IEEE 802.1w

Blind Maximum-Likelihood Estimation of Traffic Matrices in Long Range Dependent Traffic

Pier Luigi Conti[1], Livia De Giovanni[2], and Maurizio Naldi[3]

[1] Università di Roma "La Sapienza", Dip. di Statistica, Probabilità e Statistiche Applicate, Piazzale Aldo Moro, Rome, Italy
[2] Università LUMSA, Piazza delle Vaschette 101, Rome, Italy
[3] Università di Roma "Tor Vergata", Dip. di Informatica, Sistemi e Produzione, Via del Politecnico 1, Rome, Italy

Abstract. A method is proposed to estimate traffic matrices in the presence of long-range dependent traffic, while the methods proposed so far for that task have been designed for short-range dependent traffic. The method employs the traffic measurements on links and provides the maximum likelihood estimate of both the traffic matrix and the Hurst parameter. It is "blind", i.e. it does not exploit any model neither for the traffic intensity values (e.g. the gravity model) nor for the mean-variance relationship (e.g. the power-law model). In the application to a sample network the error on traffic intensities decays rapidly with the traffic intensity down to below 30%. The estimation error of the Hurst parameter can be reduced to a few percentage points with a proper choice of the measurement interval.

1 Introduction

Traffic matrices represent an essential information for network design. They contain the traffic intensity values for the OD (Origin-Destination) pairs within the network [1]. Unfortunately, the direct measurement of those values is quite burdensome and the matrix is generally estimated on the basis of traffic measurements on links, usually available. However, in a network with V nodes, the number of OD pairs (i.e. our unknowns in th estimation task) grows as their square while the number of links (i.e. the available data) is typically proportional to V. We have then an incomplete information problem. The traffic matrix estimation problem has been dealt with in the past for circuit-switched networks (see [2], [3], [4]) and, more recently, for packet data networks (see [5], [6], [7], [8], [9]). However, all the studies so far have neglected the correlation between the traffic for the same OD pair at different times. A number of papers have shown that the IP traffic exhibits time correlation even over long timescales and can be modelled as a long-range dependent (LRD) traffic (see e.g. [10], [11], and [12]). In this context the methods based on the Poisson assumption lose their effectiveness. In this paper we propose a method to estimate traffic matrices when the underlying traffic process is long-range dependent and provide an early application of the method to a sample network.

R. Valadas and P. Salvador (Eds.): FITraMEn 2008, LNCS 5464, pp. 141–154, 2009.
© Springer-Verlag Berlin Heidelberg 2009

2   Traffic Matrices in Long-Range-Dependent Traffic

We consider a network composed of V nodes; the resulting Origin-Destination (OD) pairs are $N = V(V-1)/2$. The ends of each OD pair are connected by one of M transmission links. In general the topology is not a full mesh so that $M < N$. The traffic flows from any origin to any destination, and its intensity at time t is represented by the random vector $\underline{X}^{(t)} = \left(X_1^{(t)} X_2^{(t)} \ldots X_N^{(t)} \right)$, composed by N independent elements. We suppose that the traffic is stationary (at least during our measurement period), so that both its expected value and its variance (μ_{X_i} and $\sigma^2_{X_i}$ for the i-th pair, respectively) do not change in time. We can gather the expected values pertaining to the N pairs in the vector

$$\underline{\mu}_X = \begin{pmatrix} \mu_{X_1} \\ \mu_{X_2} \\ \ldots \\ \mu_{X_N} \end{pmatrix}. \tag{1}$$

Though the traffic generated by each node is independent of any other node, each OD traffic is a long range dependent process, so that its autocorrelation decays according to a polynomial law. For any OD pair the autocorrelation matrix is

$$R = \begin{pmatrix} 1 & \rho(1) & \rho(2) & \ldots \rho(T-1) \\ \rho(1) & 1 & \rho(1) & \ldots \rho(T-2) \\ \rho(2) & \rho(1) & 1 & \ldots \rho(T-3) \\ \ldots & \ldots & \ldots & \ldots & \ldots \\ \rho(T-1) & \rho(T-2) & \rho(T-3) & \ldots & 1 \end{pmatrix}, \tag{2}$$

where

$$\rho(k) = \frac{1}{2} \left[(k+1)^{2H} - 2k^{2H} + (k-1)^{2H} \right], \tag{3}$$

and H is the Hurst parameter. In the sequel we assume that the Hurst parameter is the same for each OD pair. While the autocorrelation matrix is the same for all the OD couples, the autocovariance matrix depends instead on the variance of each OD pair traffic:

$$\Sigma_{X_i} = \sigma^2_{X_i} R. \tag{4}$$

The two assumptions of independence and stationarity are well supported in the literature.

The assumption of independence between OD pairs has been studied for real data in [13], via the correlation between the standardized residuals of the bits (packets) arrival process (bit/packet network traffic) of two different OD pairs at various time aggregation levels. The data at hand are measurements observed on one link in the Finnish university network (Funet), and partitioned into OD traffic on the basis of origin and destination IP address; time aggregation varies from 1 sec to 300 secs. The analysis of the empirical distribution of the correlation coefficients and of their greatest values does not show any particular evidence against the assumption of independence between OD pairs. A proper statistical test for each OD pair should be conducted separately.

The assumption of independence within OD pairs has been recently considered at time scales either relevant for traffic matrix estimation or suitable to detect long memory (time aggregation varying from 1 sec to 300 secs). In particular, the presence of long range dependence of traffic data is shown in [13],[15] through the analysis of the correlation between the standardized residuals of the bits (packets) arrival process at lag k of an OD pair at various time aggregation levels; in [12] through wavelet analysis (with a time aggregation for 10 msecs to 60 secs), in [14] through a visual analysis. The above mentioned papers show that OD traffic is characterized by the presence of long range dependence. As to the persistence of long range dependence, it has been observed in [12] that a significant correlation is present for periods over half an hour, even when the aggregation measurement intervals are 5 minutes long. The homogeneity hypothesis for H relies on the consideration that long memory of network traffic is due to the multiple time scale nature of traffic coupled with transport protocol aspects [11]. Assuming that OD traffic is the superposition of sources with different H values, according to large deviations theory sources with highest H will then dominate.

The assumption of stationarity is studied in [6], [14]. In [6] the time varying nature of network traffic is visually observed in Lucent data (traffic observed on the links and OD pairs of a one-router network with 5 minutes time aggregation as provided by SNMP, the Simple Network Management Protocol). Stationarity of OD traffic (w.r.t. mean and variance) is assumed to hold in a window lasting up to 21 five minutes time-intervals, i.e. slightly less than 2 hours. To take into account the time-varying nature of OD traffic, in [6] it is proposed to estimate OD traffic (via EM) using a local i.i.d model within a moving data window. In [14] the assumption of stationarity is considered via the correlation coefficient between packet network traffic in adjacent periods (time aggregation from milliseconds to seconds). The main conclusion is that stationarity can be reasonably assumed to hold within a period of 30-90 minutes. The empirical studies performed in [15], [13], [16] for Funet data essentially confirm the stationarity assumption by [6] (again, a time aggregation from 1 sec to 300 secs is used). The stationarity period allows one to gather a sufficiently large set of measurements, if they are taken at high frequency: a measurement frequency of 1 minute (anyway finer than the 5 minute resolution provided by SNMP) would result in 30 measurements over half an hour.

In this context the origin-destination traffic is supposed not to be observable. The only observable quantities are the traffic intensity values as measured on the transmission links. Similarly as above we define the vector representing the traffic process as observed on the transmission links $\underline{Y}^{(t)} = \left(Y_1^{(t)} Y_2^{(t)} \dots Y_M^{(t)} \right)$, whose size is M. This vector is related to the OD traffic by the relationship

$$\underline{Y}^{(t)} = A \cdot \underline{X}^{(t)}, \tag{5}$$

where the matrix A, of size MxN, is the routing matrix, whose element a_{kl} is 1 if the k-th transmission link carries the traffic pertaining to the l-th OD pair. Being a linear combination of OD traffic components, each link traffic component possesses an expected value constant over time

$$\underline{\mu}_Y = E[\underline{Y}^{(t)}] = AE[\underline{X}^{(t)}] = A \begin{pmatrix} \mu_{X_1} \\ \mu_{X_2} \\ \cdots \\ \mu_{X_N} \end{pmatrix} = A\underline{\mu}_X. \tag{6}$$

Since the matrix A does have a rank smaller than N (usually M), the model introduced so far is unidentifiable. In order to make it identifiable, a mean-variance relationship is usually assumed in the literature (where only short-range dependence models are considered). In the present paper, we avoid such an assumption.

3   Joint Statistics of Observables and Origin-Destination Traffic

In this section, as a step towards the definition of the maximum likelihood estimator we derive some joint statistics of interest. Though the traffic originated by each OD pair is independent of all other OD pairs, it is correlated with the traffic observed on transmission links, namely with those transmission links that carry it. In order to determine this correlation, we define a column vector $\underline{Z}_i$ for each OD pair, which gathers both the traffic pertaining to the i-th OD pair and the traffic carried by the transmission links:

$$\underline{Z}_i = \begin{pmatrix} \underline{X}_i \\ \underline{Y}^{(1)} \\ \cdots \\ \underline{Y}^{(T)} \end{pmatrix} = \begin{pmatrix} \underline{X}_i \\ \underline{Y} \end{pmatrix}. \tag{7}$$

Such random vector possesses multinormal distribution with $(M+1)T$ components. Each component of $\underline{Z}_i$ is correlated with the other components. The resulting covariance matrix can be written as:

$$M_{Z_i} = \begin{pmatrix} \Sigma_{X_i X_i} & \Sigma_{X_i Y} \\ \Sigma_{Y X_i} & \Sigma_{YY} \end{pmatrix}, \tag{8}$$

where (using the form $\mathrm{Cov}(\underline{A}, \underline{B})$ to denote the covariance matrix of vectors $\underline{A}$ and $\underline{B}$), the submatrices are defined as

$$\begin{aligned} \Sigma_{X_i X_i} &= \mathrm{Cov}(\underline{X}_i, \underline{X}_i), \\ \Sigma_{X_i Y} &= \mathrm{Cov}(\underline{X}_i, \underline{Y}), \\ \Sigma_{Y X_i} &= \mathrm{Cov}(\underline{Y}, \underline{X}_i) = \Sigma'_{X_i Y}, \\ \Sigma_{YY} &= \mathrm{Cov}(\underline{Y}, \underline{Y}). \end{aligned} \tag{9}$$

The expressions introduced so far may be reformulated, in order to obtain simpler expressions. Consider first the mixed covariance submatrix, whose expression is

$$\Sigma_{X_iY} = E\left[\left(\underline{X}_i - \underline{\mu}_{X_i}\right)\left(\underline{Y} - \underline{\mu}_Y\right)'\right]$$
$$= \left[P_i^{(1)}A' \; P_i^{(2)}A' \; \cdots \; P_i^{(T)}A'\right], \tag{10}$$

where $\underline{\mu}_{X_i} = \mu_{X_i}\underline{1}$, and $P_i^{(s)}$ $(s = 1, 2, \ldots, T)$ are the TxN matrices defined herafter, whose only non zero column is the i-th one

$$P_i^{(s)} = E\left[\left(\underline{X}_i - \underline{\mu}_{X_i}\right)\left(\underline{X}^{(s)} - \underline{\mu}_X\right)'\right]$$
$$= \begin{bmatrix} 0 & \cdots & 0 & \sigma_{X_i}^2\rho(s-1) & 0 & \cdots & 0 \\ 0 & \cdots & 0 & \sigma_{X_i}^2\rho(s-2) & 0 & \cdots & 0 \\ \cdots\cdots\cdots & & & \cdots & & \cdots\cdots\cdots \\ 0 & \cdots & 0 & \sigma_{X_i}^2\rho(s-T) & 0 & \cdots & 0 \end{bmatrix}. \tag{11}$$

By introducing the matrices describing the correlation between any two different OD pairs at time instants i and j

$$\Sigma_{X^{(i)}X^{(j)}} = E\left[\left(\underline{X}^{(i)} - \underline{\mu}_X\right)\left(\underline{X}^{(j)} - \underline{\mu}_X\right)'\right]$$
$$= \begin{bmatrix} \sigma_{X_1}^2\rho(i-j) & 0 & & 0 \\ 0 & \cdots & & 0 \\ 0 & \cdots & & 0 \\ 0 & & 0 & \sigma_{X_N}^2\rho(i-j) \end{bmatrix} \tag{12}$$
$$\equiv \Sigma_{|i-j|},$$

the covariance matrix of the observables is similarly expressed in the compact form

$$\Sigma_{YY} = E\left[\left(\underline{Y} - \underline{\mu}_Y\right)\left(\underline{Y} - \underline{\mu}_Y\right)'\right]$$
$$= \begin{pmatrix} A\Sigma_0 A' & A\Sigma_1 A' & \cdots & A\Sigma_{T-1}A' \\ A\Sigma_1 A' & A\Sigma_0 A' & \cdots & A\Sigma_{T-2}A' \\ \cdots & \cdots & \cdots & \cdots \\ A\Sigma_{T-1}A' & A\Sigma_{T-2}A' & \cdots & A\Sigma_0 A' \end{pmatrix} \tag{13}$$

4   Statistics of OD Traffic Conditioned to the Observables

After having shown in Section 3 that Z_i has a multinormal distribution with a covariance matrix M_{Z_i}, we now turn to the OD traffic statistics. Though they are not directly observable, we can use the traffic as measured on the transmission links to update their estimates. In particular, the expected value of the OD traffic conditioned to the observables,

$$m_{X_i} \equiv E\left[\underline{X}_i | \underline{Y}^{(1)}, \dots, \underline{Y}^{(T)}, \underline{\theta}\right], \qquad (14)$$

and the covariance matrix

$$\Sigma_{X_i}^* \equiv E\left[\left(\underline{X}_i - \underline{m}_{X_i}\right)\left(\underline{X}_i - \underline{m}_{X_i}\right)' | \underline{Y}^{(1)}, \dots, \underline{Y}^{(T)}, \underline{\theta}\right] \qquad (15)$$

are of interest. Here $\underline{\theta} = \left(\mu_{X_1}, \dots, \mu_{X_N}, \sigma_{X_1}^2, \dots, \sigma_{X_N}^2, H\right)$ denotes the parameter vector of the probability density function of the vector traffic variable (as made clear in the subsequent sections, they will be computed using the latest estimate of $\underline{\theta}$). The distribution of $\underline{X}_i$ conditioned to the observables is still multinormal, with means and covariances

$$\underline{m}_{X_i} = \underline{\mu}_{X_i} + \Sigma_{X_i Y} \Sigma_{YY}^{-1} \left(\underline{Y} - \underline{\mu}_Y\right) \qquad (16)$$

$$\Sigma_{X_i}^* = \Sigma_{X_i X_i} - \Sigma_{X_i Y} \Sigma_{YY}^{-1} \Sigma_{Y X_i}. \qquad (17)$$

5   EM Estimation of Traffic Matrices

We are interested in obtaining at the same time an estimate of the traffic intensity value for all the OD pairs and the common value of the Hurst parameter H. In the expressions reported so far, the unknown quantities are the elements of the vector $\underline{\theta} = \left(\mu_{X_1}, \dots, \mu_{X_N}, \sigma_{X_1}^2, \dots, \sigma_{X_N}^2, H\right)$. Our goal is to obtain an estimate of this vector by the maximum likelihood function method. In this section we provide first the expression of such function, and then the description of the method employed to obtain our estimates of $\underline{\theta}$.

Since the traffic generated by each OD pair at any given time is independent of the traffic generated by any other OD pair, we can write the likelihood function in a product form, where each term is the likelihood function pertaining to a sample of T elements extracted from a multinormal distribution (pertaining to a single OD pair):

$$L_X(\theta) = \prod_{i=1}^{N} \frac{1}{\sqrt{(2\pi\sigma_{X_i}^2)^T |R|}} \exp\left[-\frac{d_i' R^{-1} d_i}{2\sigma_{X_i}^2}\right], \qquad (18)$$

where $d_i = \underline{x}_i - \underline{\mu}_{X_i}$. In order to maximize this function we resort to the Expectation-Maximization (EM) algorithm [17] [18]. This iterative algorithm provides at each iteration step a refined estimate of $\underline{\theta}$; namely at the t-th step it provides the new estimate

$$\underline{\theta}_t = \arg\max_{\theta} \mathcal{L}(\underline{\theta}_{t-1}, \underline{\theta}) \qquad (19)$$

where we use the logarithmic likelihood function

$$\mathcal{L}(\underline{\theta}_{t-1}, \underline{\theta}) = E\left\{\ln\left[L_X(\underline{\theta})\right] | Y^{(1)}, \dots, Y^{(T)}; \underline{\theta}_{t-1}\right\}. \qquad (20)$$

The obtained estimators are $\sqrt{n}$-consistent, with $n = T \cdot N$.

Although the model itself is not identifiable, EM algorithm (without any mean-variance relationship assumption) still produces maximum likelihood estimates of the parameters of interest [19]. The EM algorithm consists of the following two steps.

1. Computation of the expected value of the likelihood function conditioned to the observables (Expectation step);
2. Maximization of the likelihood function with respect to the unknown parameters (Maximization step).

As to the Expectation step, by fully developing the definition (20) we obtain the following expression

$$
\mathcal{L}_X \left(\underline{\theta}_{t-1}, \underline{\theta} \right) = E \left\{ \ln \left[L_X \left(\underline{\theta} \right) \right] | y^{(1)}, \ldots, y^{(T)}; \underline{\theta}_{t-1} \right\}
$$

$$
= \sum_{i=1}^{N} \left[-\frac{T}{2} \ln(2\pi) - T \ln \sigma_{X_i} - \frac{1}{2} \ln |R| \right]
$$

$$
- \sum_{i=1}^{N} E \left\{ \frac{\left(\underline{x}_i - \underline{\mu}_{X_i} \right)' \Sigma_{X_i}^{-1} \left(\underline{x}_i - \underline{\mu}_{X_i} \right)}{2} \Big| y^{(1)}, \ldots, y^{(T)}; \underline{\theta}_{t-1} \right\}.
$$

$$(21)$$

Then, by using the commutative property for the trace of the product of two matrices (as already exploited in a similar context in [20]), and by some algebraic manipulation, the expression

$$
\mathcal{L}_X (\underline{\theta}_{t-1}, \underline{\theta}) = \sum_{i=1}^{N} \left[-\frac{T}{2} \ln (2\pi) - T \ln \sigma_{X_i} - \frac{1}{2} \ln |R| \right]
$$

$$
- \sum_{i=1}^{N} \frac{1}{2\sigma_{X_i}^2} \mathrm{Tr} \left\{ R^{-1} \Sigma_{X_i}^* \right\}
$$

$$(22)$$

$$
- \sum_{i=1}^{N} \frac{1}{2\sigma_{X_i}^2} \mathrm{Tr} \left\{ \left(\underline{m}_{X_i} - \underline{\mu}_{X_i} \right)' R^{-1} \left(\underline{m}_{X_i} - \underline{\mu}_{X_i} \right) \right\}.
$$

is obtained. We now turn to the problem of estimating the means and variance of traffic. Since the stochastic process $\underline{X}_i = \left(\underline{X}^{(1)}, \underline{X}^{(2)}, \ldots, \underline{X}^{(T)} \right)$ is stationary, the vector $\underline{\mu}_{X_i}$ is formed by identical elements. We can then differentiate w.r.t. its components and take the derivatives equal to zero to obtain the estimates

$$
\mu_{X_i} = \frac{\underline{m}'_{X_i} \Sigma_{X_i}^{-1} \underline{1}}{\underline{1}' \Sigma_{X_i}^{-1} \underline{1}} = \frac{\underline{m}'_{X_i} R^{-1} \underline{1}}{\underline{1}' R^{-1} \underline{1}} \qquad \forall i = 1, .., N. \tag{23}
$$

The variances are similarly estimates, i.e. again by differentiating the log-likelihood function nad by taking its derivatives equal to zero. In this way, the estimates

$$
\sigma_{X_i}^2 = \frac{1}{T} \left[\mathrm{Tr} \left(R^{-1} \Sigma_{X_i}^* \right) + \left(\underline{m}_{X_i} - \underline{\mu}_{X_i} \right)' R^{-1} \left(\underline{m}_{X_i} - \underline{\mu}_{X_i} \right) \right]. \tag{24}
$$

are obtained. By replacing these expressions in the likelihood function (22), the quantities $\underline{m}_{X_i}$, $\underline{\mu}_{X_i}$ and $\sigma^2_{X_i}$ being computed by using the latest estimate of the Hurst parameter, the expected value of the log-likelihood function turns out to be a function of the autocorrelation matrix R, which in turn depends on the Hurst parameter only. Hence the optimal value of the Hurst parameter is obtained by a single-unknown maximization procedure. Though we don't provide a formal proof, in all the cases examined the likelihood function is a smooth quasi-parabolic function of the Hurst parameter, with a global maximum (i.e. a unique solution to the maximization problem). For the maximization task we have employed Brent's method [21].

6   Simulation Results

In order to evaluate the performance of our methods we have performed a simulation on a small-scale network. It is to be stressed that, in our knowledge, no methods to estimate traffic matrices in the context of long range dependence have been proposed so far. Hence, there are no previous research results that can be used as a benchmark. We provide a comparison with the results obtained by methods that consider the quite simple scenario of short-range-dependent processes. As figures of merit we employ the following quantities

- Error on the Hurst parameter;
- Error on traffic intensities.

In the following subsections we first describe the network considered in our simulation and the characteristic of the traffic with which we load it, and then provide the simulation results.

6.1   Network Model

In order to evaluate the performance of the traffic estimation algorithm presented in the previous sections we consider a toy network, already used by Vardi in his seminal paper [5]. This network, reported in Fig. 1 where the nodes are labelled by letters and the links by numbers, is made of 4 nodes (hence 12 origin-destination relationships), and 7 unidirectional links. The routing matrix (where the OD pairs listed on the columns are sorted in lexicographic order) is

$$
\mathbf{A} = \begin{pmatrix}
1 & 0 & 0 & 0 & 0 & 0 & 0 & 0 & 0 & 0 & 0 & 0 \\
0 & 1 & 1 & 0 & 0 & 1 & 0 & 0 & 0 & 0 & 0 & 0 \\
0 & 0 & 0 & 1 & 0 & 1 & 1 & 0 & 0 & 1 & 0 & 0 \\
0 & 0 & 0 & 0 & 1 & 0 & 0 & 0 & 0 & 0 & 0 & 0 \\
0 & 0 & 0 & 0 & 0 & 0 & 1 & 1 & 0 & 1 & 1 & 0 \\
0 & 0 & 1 & 0 & 0 & 1 & 0 & 0 & 1 & 0 & 0 & 0 \\
0 & 0 & 0 & 0 & 0 & 0 & 0 & 0 & 0 & 1 & 1 & 1 \\
\end{pmatrix} \tag{25}
$$

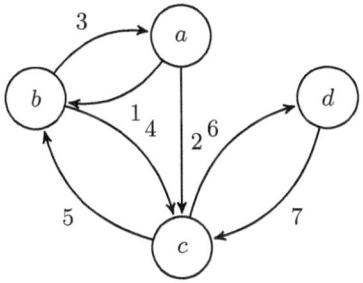

Fig. 1. Toy network

6.2   Traffic Model

In order to simulate the traffic measured on the links and apply our matrix estimation algorithm, we make the following assumptions on the traffic generated.

1. The traffic generated by each node is assumed to follow a long range dependence process, so that the autocorrelation functions obeys the relationship (3);
2. The stochastic processes associated to any two OD pairs are stochastically independent of each other;
3. All the OD pairs have the same value of the Hurst parameter;
4. The expected value of traffic follows a Zipf rank-size relationship;
5. The mean-variance relationship for the traffic intensity follows a power law;
6. The coefficient of preference, which determines how the traffic generated by a given origin node distributes among all the destinations, is proportional to the traffic generated by the destination node.

It is to be noted that only Assumptions (1)-(3) are actually used in the proposed estimation method. Assumptions (4)-(6) are just used for the purpose of generating the synthetic traffic data employed to feed the estimation algorithm and could be replaced by different ones (though, as explained in the following, they have been chosen as supported by the literature on the subject); they are not exploited in the estimation algorithm.

We now proceed to explain and justify each of these assumptions. Assumptions (1) through (3) are essential to the algorithm proposed for traffic matrix estimation. Assumptions (4) through (6) are instead employed to set our simulation in a realistic framework, but are not exploited in any step of the estimation process.

Assumptions (1) and (2) have been shown in Section 2 to be well supported in the literature. Assumption (3) is due to the fact that we do not expect the set of applications run by customers connected to a given node to differ significantly from those used by the customers of a different node, or, in other words, nodes (representing routers) are not specialized by service. Though we do not know of any thorough work devoted to the relation between the Hurst parameter and the network application, it has since long been known that the feedback behaviour built in the TCP transport protocol (and hence in all the network applicaiton relying on it) is a determinant of long range dependence. More recently, it has been observed that the appearance of a particular P2P protocol (namely Blubster, a.k.a. Piolet) has been related to a variation of the measured Hurst parameter [12]. Hence we can expect the Hurst parameter being somewhat determined by the mix of applications generating the observed traffic. Therefore it should not be different among nodes generating roughly the same application mix. Assumption (4) means that, if we sort the nodes by their average generated traffic intensity in decreasing order, we expect that intensity to be related to the node rank (rank 1 corresponding the maximum traffic node) by the Zipf law

$$\mu_{O_{(i)}} \propto \frac{1}{i^\alpha} \qquad i = 1\ldots,V. \qquad (26)$$

This law, originally formulated in the context of linguistics [22], is found in many different contexts to describe rank-frequency relationships. The parameter α governs the imbalance of the traffic distribution: the larger α, the larger the differences in the traffic intensity between the highest and the lowest ranked nodes. As to its applications in a telecommunications environment, Zipf law is supported by measurements conducted on the telephone network and on Internet users [23]. Assumption (5) is expressed by the relationship

$$\sigma_{X_i}^2 = \lambda \cdot \mu_{X_i}^c \qquad i = 1,\ldots,N. \qquad (27)$$

It was put forward in the seminal paper by Cao [6] and is supported by several measurements campaigns [8] [24] [25]. Assumption (6) is common in teletraffic studies, e.g. Chapter 13 in the reference book [1]. If we indicate by μ_{O_k} the traffic intensity generated by the k-th node, we can then associate each OD pair to the endnodes forming that pair. For the generic OD pair with traffic intensity X_i, we can denote its origin node as O_l and its destination node as O_m. Using this position, the expected traffic intensity for that OD pair is

$$\mu_{X_i} = \mu_{O_l} \frac{\mu_{O_m}}{\sum_{k=1}^{V} \mu_{O_k}} \qquad i = 1,\ldots,N. \qquad (28)$$

Aa a direct consequence of the combination of Assumptions (4) and (6), the resulting matrix of expected traffic intensities is asymmetric.

Finally, the long-range-dependent traffic traces are generated by using the Choleski method [26]. This method, though computationally heavy, is exact and has been become the reference method for such task [27].

6.3   Simulation Parameters

In our simulation we have employed the following set of parameter values

  − H=0.6, 0.8;
  − Zipf parameter $\alpha = 1$;

- Number of samples $T = 30, 90$;
- Parameter of the power-law relationship between mean and variance $\lambda = 1$ and $c = 1$.

The stopping criterion for the search of the optimal Hurst parameter has been based on the relative updating step: the search is stopped when the new estimate differs less than 2% from the previous estimate. Though this may somewhat reduce the accuracy of the results, it also reduces the simulation time to reasonable values, considering the mass of simulation runs dealt with in this first application.

6.4   Hurst Parameter Estimation

The first quantity of interest is the Hurst parameter. Though this parameter is not the main goal of our estimating task (we are trying to estimate the traffic matrix), the capability to provide reliable estimates of H is an important measure of how well our estimating procedure captures the long-range-dependent nature of the process at hand. The results obtained for the Hurst parameter are shown in Table 1. The values reported are the mean values of the estimates as gathered over a block of 500 simulation runs. We can observe a slight underestimation of that parameter in all cases. Estimates are better for the lower values of the Hurst parameter. However, the bias is progressively reduced when longer traffic traces are used. Trebling the trace length reduces the relative error roughly by a factor of four when $H = 0.6$ (from 9.8% to 2.3%) and halves it when $H = 0.8$ (from 11.1% to 5.4%).

Table 1. Estimated Hurst parameter

	H=0.6	H=0.8
T=30	0.541	0.711
T=90	0.586	0.757

6.5   Traffic Intensity Estimation

We now turn to the main goal, i.e. the estimation of the traffic matrix itself. Here we report the relative estimation error for the cases listed in Section 6.3 as resulting from the average of the errors occurring over a block of 500 simulation runs. Rather than reporting the average error over all the OD pairs we perform the analysis for each OD pair separately. In fact, our aim is to obtain low relative errors especially for the largest traffic OD pairs. That line of reasoning suggest to focus on the largest traffic OD pairs, e.g. on those making up 90% of the whole traffic [9]. In Fig. 2 we report the error as a function of the actual traffic intensity value. As desired the error decays rapidly with the traffic intensity, achieving values as low as 25%. The average error over all the OD pairs is in the 59%-65% range. The OD pairs making up 90% of the overall traffic volume are

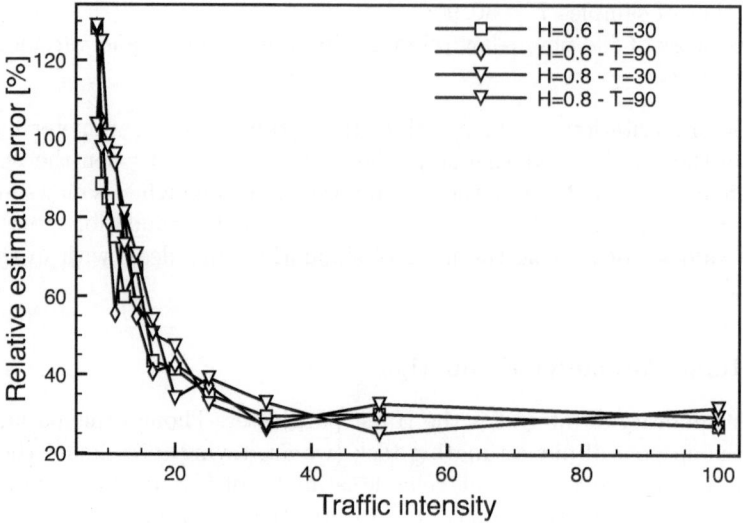

Fig. 2. Estimation error on traffic intensity

those whose traffic intensity is above 11. For those OD pairs the average error is instead in the 45%-51% range. Better estimates are in both cases achieved for the lowest values of the Hurst parameter. The trace length seems to have a negligible effect on the quality of estimates. As previously recalled we lack a benchmark to compare these figure against, since no other traffic matrix estimation methods have been proposed for LRD traffic. However, the figures reported in [9] for the overall average error in short-range dependent traffic for the three methods there considered (projection method, constrained optimization, and maximum likelihood estimation) lie in the 29%-110% range.

7   Conclusion

We have proposed a method for the estimation of the traffic matrix in the case of long-range dependent traffic, based on the maximum likelihood principle and on the numerical EM procedure. As a by-product the method also provides the estimate of the Hurst parameter. The method has been applied to the same toy network considered in previous seminal papers. The estimate of the Hurst parameter is slightly negatively biased, but such bias can be reduced to a few percentage points by using longer traffic traces. The estimates of traffic intensity are better for the largest traffic OD pairs, for which the relative error gets below 30%. Though no direct benchmarks exist, the errors lie within the range obtained with previous techniques for short-range dependent traffic. Both the estimates (of the Hurst parameter and of the traffic intensity) are better for the lowest values of the Hurst parameter. It is to be noted that, contrary to most of such techniques, the method here proposed is completely blind, i.e. doesn't exploit any model either for the traffic intensity values (e.g. the gravity model) or for the mean-variance relationship (e.g. the power-law model).

Acknowledgment

The authors wish to acknowledge the contributions of the MIUR PRIN projects MAINSTREAM (2006) and "Analisi di problemi complessi in presenza di informazione incompleta: metodologie statistiche e applicazioni" (2005), and of the Euro-NF Network of Excellence, funded by the European Union.

References

1. Bear, D.: Principles of telecommunication traffic engineering. IEE/Peter Peregrinus, London (1988)
2. Kim, N.: A point-to-point traffic estimation from trunk-group and office measurements for large networks. In: 13th International Teletraffic Congress, Copenhagen, Denmark (June 1991)
3. Tu, M.: Estimation of point-to-point traffic demand in the public switched telephone network. IEEE Transactions on Communications 42(2/3/4), 840–845 (1994)
4. Buttó, M., Conversi, P.G., Naldi, M.: Estimation of point-to-point traffic in circuit-switched networks. European Transactions on Telecommunications 10(5), 497–504 (1999)
5. Vardi, Y.: Network tomography: Estimating source-destination traffic intensities from link data. Journal of the American Statistical Association 91, 365–377 (1996)
6. Cao, J., Davis, D., Vander Wiel, S., Yu, B.: Time-varying network tomography: Router link data. Journal of the American Statistical Association 95, 1063–1075 (2000)
7. Castro, R., Coates, M., Liang, G., Nowak, R., Yu, B.: Network tomography: Recent developments. Statistical Science 19, 499–517 (2004)
8. Medina, A., Taft, N., Salamatian, K., Bhattacharyya, S., Diot, C.: Traffic matrix estimation: Existing techniques and new directions. In: SIGCOMM 2002, Pittsburgh, Pennsylvania, USA (August 2002)
9. Juva, I.: Traffic matrix estimation in the Internet: measurement analysis, estimation methods and applications. PhD in Technology, Helsinki University of Technology (2007)
10. Karagiannis, T., Molle, M., Faloutsos, M.: Long-range dependence: Ten years of internet traffic modeling. IEEE Internet Computing 8(5), 57–64 (2004)
11. Gong, W.B., Liu, Y., Misra, V., Towsley, D.: Self-similarity and long range dependence on the internet: a second look at the evidence, origins and implications. Computer Networks 48(3), 377–399 (2005)
12. Park, C., Hernandez-Campos, F., Marron, J.S., Donelson Smith, F.: Long-range dependence in a changing internet traffic mix. Computer Networks 48(3), 401–422 (2005)
13. Susitaival, R., Juva, I., Peuhkuri, M., Aalto, S.: Characteristics of origin-destination pair traffic in funet. Telecommunication Systems 33, 67–88 (2006)
14. Norros, I., Kilpi, J.: Testing the gaussian approximation of aggregate traffic. In: Internet Measurement Workshop IMW 2002, Marseille (2002)

15. Juva, I., Susitaival, R., Peuhkuri, M., Aalto, S.: Traffic characterization for traffic engineering purposes: analysis of funet data. In: Proceedings of the 1st EuroNGI Conference on Next Generation Internet Networks (NGI 2005), Rome, pp. 404–422 (2005)
16. Juva, I., Susitaival, R., Peuhkuri, M., Aalto, S.: Effects of spatial aggregation on the characteristics of origin-destination pair traffic in funet. In: Koucheryavy, Y., Harju, J., Sayenko, A. (eds.) NEW2AN 2007. LNCS, vol. 4712, pp. 1–12. Springer, Heidelberg (2007)
17. Dempster, A.P., Laird, N.M., Rubin, D.B.: Maximum likelihood from incomplete data via the EM algorithm. Journal of the Royal Statistical Society, Series B (Methodological) 39(1), 1–38 (1977)
18. Moon, T.K.: The expectation-maximization algorithm. IEEE Signal Processing Magazine 13(6), 47–60 (1996)
19. Kuk, A., Chan, J.: Three ways of implementing the EM algorithm when parameters are not identifiable. Biometrical Journal 43, 207–218 (2001)
20. Juva, I., Vaton, S., Virtamo, J.: Quick traffic matrix estimation based on link count covariances. In: 2nd EuroNGI Workshop on Traffic Engineering, Protection and Restoration for NGI, Rome, Italy, April 21-22 (2005)
21. Press, W., Teukolsky, S., Vetterling, W., Flannery, B.: Numerical Recipes in C, 2nd edn. Cambridge University Press, Cambridge (1992)
22. Zipf, G.K.: Human Behavior and the Principle of Least Effort. Addison-Wesley, New York (1949)
23. Naldi, M., Salaris, C.: Rank-size distribution of teletraffic and customers over a wide area network. European Transactions on Telecommunications 17(4), 415–421 (2006)
24. Susitaival, R., Juva, I., Peuhkuri, M., Aalto, S.: Characteristics of origin-destination pair traffic in funet. Telecommunication Systems 33(1-3), 67–88 (2006)
25. Gunnar, A., Johansson, M., Telkamp, T.: Traffic matrix estimation on a large ip backbone: a comparison on real data. In: IMC 2004: Proceedings of the 4th ACM SIGCOMM conference on Internet measurement, pp. 149–160. ACM, New York (2004)
26. Hall, P., Jing, B., Lahiri, S.N.: On the sampling window method for long-range dependent data. Statistica Sinica 8(4), 1189–1204 (1998)
27. De Giovanni, L., Naldi, M.: Tests of correlation among wavelet-based estimates for long memory processes. Communications in Statistics - Simulation and Computation 37(2), 301–313 (2008)

Optimizing Network Performance in Multihoming Environments

Nuno Coutinho, Susana Sargento, and Vitor Jesus

Instituto de Telecomunicações, Universidade de Aveiro, Portugal
{nunocoutinho,susana}@ua.pt, vjesus@av.it.pt

Abstract. Future Internet will be based in heterogeneous infrastructures with a large landscape of access technologies. Moreover, the terminals are being equipped with multiple interfaces, enabling the simultaneous connection to the different available technologies. In such multihoming environments, the traffic control mechanisms will require complex operations, as it is now possible to implement the concept of "always best connected" (ABC), using the devices and access technologies that best suit communications needs, users and networks. This paper presents a performance study of a traffic control mechanism that is able to perform context-aware and personalized network selection, using context and preferences, network and terminal characteristics, to determine the best access connection for each terminal and service. The results show the benefits of using such an algorithm in the network, mainly in terms of QoS in these multiservice technologies, and address the influence of specific criteria and constraints considered in the decision process.

Keywords: Multihoming, always best connected, context-awareness, intelligent decisions, mobility.

1 Introduction

Various access technologies, such as WiFi, UMTS, DVB, HSDPA and WiMAX, are now available to mobile devices, which are increasingly equipped with more interfaces of different technologies (multihomed). Due to these improvements, the next generation of mobile communications will be based on a heterogeneous landscape of different technologies that can be combined in a common platform to provide complementary services. In this environment each mobile terminal will be able to connect simultaneously to different technologies, which vary in bandwidth, delay, communication range, power consumption, security, reliability, cost and several other aspects. Therefore, the concept of being always connected may change to the concept of always best connected (ABC) [1], enabling the choice of the best point of attachment to each user/services.

However, the definition of best may have different perspectives depending on several aspects that may be subjective or objective. Characteristics like personal preferences, device capabilities, application requirements, network coverage, resources and network load are strictly related with best connectivity. This paper discusses this

R. Valadas and P. Salvador (Eds.): FITraMEn 2008, LNCS 5464, pp. 155–168, 2009.
© Springer-Verlag Berlin Heidelberg 2009

multiple and arbitrary criteria problem at the mobility layer of networks. From a top-level view, the problem considers the mapping of a flow to an access point, considering not only available resources (e.g., for QoS concerns) but also the circumstances that triggered the decision algorithm.

This paper focus on the access selection process to optimize network performance when considering multi-technology and multihoming environments, using any-constraint algorithm based on parameters related to context, preferences, and terminal and network characteristics, combining this knowledge to enable the optimization of both terminal and network point of view [3]. In order to evaluate the efficiency of the implemented algorithm, different scenarios were simulated showing the benefits of the selection scheme. The simulations focus on the performance evaluation of the mechanism and the impact of its parameters in the decision process, in order to highlight the criteria flexibility, functionality and efficiency of the scheme.

The rest of the paper has the following structure: section 2 briefly discusses other proposals of network selection algorithms. Section 3 conceptually describes the algorithm implemented and its main features, and section 4 presents the results of the network selection process. Finally, section 5 concludes the paper and introduces topics for further research.

2   Related Work

In this section, we review related work, setting our requirements and briefly compare selected related work.

Some work has already been done which explores the described optimization problem. The most common approach is to center the selection process on radio signal considerations (e.g., [10]). As discussed, despite its importance, we consider that it is only one between many criteria to be accounted for network selection. It is also common to consider link quality metrics (such as delay or bandwidth) – e.g., see [11].

Song et al. [14] also use Grey analysis in a mix WLAN/UMTS environment, being the main metrics for decision the QoS desired of the user or application and the current conditions of each technology (WLAN or UMTS). We depart from this model of selection since we adopt a model of discrete services. Hence, we're closer to Gazis et al. [7] and Xing et al. [8] that model the problem of flow allocation as a knapsack problem: a user has applications to distribute across available PoAs. However, their work views mobility mainly as a resource problem, whereas we consider it to be only a part in a complete scheme for network selection.

Combined criteria not related only to signal strength or link quality have also emerged recently such as Iera et al. [5], McNair et al. [12] and Chen et al. [13]. These authors propose schemes that are based on cost functions that contribute to an overall cost function which will, in the end, determine the best (under the selected criteria) PoA to handover to.

2.1   Key Requirements and Global Perspective of Related Work

We state now the requirements that have guided our design:

(i)   *strict service admission*: clear separation between runtime admission control and quality information.

(ii) *easy plug-in of arbitrary criteria.* Any type of information should be possible to use, as long it is in a suitable format.

(iii) *flow granularity.* A flow should be the basic element.

(iv) *separation of powers.* Terminal and network should exchange information and not attempt unilateral decisions.

(v) *fast environments.* Support to queues of events.

(vi) *user optimizations.* For scalability reasons, a scheme should support local optimizations (single user) and global optimizations.

(vii) *clear separation of entities based on self-contained properties.* We have identified the following three entities whose properties should not be dependent on each other's: network infrastructure (e.g., reliability of a PoA), the user/terminal (e.g., preference for low monetary cost) and resource constraints (available resources or not in a PoA).

The review of the literature shows that, to the best of our knowledge, no scheme copes with all requirements.

3 Any-Constraint Algorithm Description

The network selection architecture implemented is based in the solution proposed in [3]. This section will briefly describe this scheme, presenting the main guidelines considered in the development of the solution, and the modeling of the several properties of each element in the network, providing an easier manipulation of the information. Finally, a description of the selection process will be given concerning triggers, the algorithm and the final handover decision made by the terminal. For detailed information, please refer to [3].

3.1 Design Guidelines

The main objective of the network selection scheme proposed is to produce a ranked list of possible handovers that the terminal is allowed to perform after any event which triggers the selection process. The ranked list is composed by flow maps [9], each containing a possible distribution of the user's flows through the available access points. This scheme addresses the several requirements listed in [9].

The events are the triggers of the architecture, caused by terminal requests, terminal movement, and any other possible change in the network that is relevant to the performance of its service provided. To support this, the scheme proposed must be able to deal with any type of trigger, being it classified as periodic, scheduled or based in context changes.

The ranked lists should be produced associated to values directly linked with QoE. However, due to the subjectivity of this metric and the difficulty of associating it with the flow maps rank, it is necessary to model the main elements in the network according to their properties.

Regarding Points of Access (PoAs), there are two obvious properties: static priorities and resources availability. Static priorities of a PoA could be reliability, monetary cost and mobility prediction. The resources of a PoA cannot be only related with bandwidth, but also with the capacity to provide different services to the user that wants to connect to it.

User properties can also be divided in static and real-time. The static properties of a user are related with all the context information that can be relevant in the handover process.

An important guideline, besides triggers and ranked lists, is that the resource management is totally independent of the ranked lists process. This means that only PoAs with resources available are allowed to enter in the flow maps calculation, making all feasible and reducing processing effort.

3.2 Entities Modeling

In order to be able to model any criteria to be used in the algorithm, we decided to format it in a matrix presentation form. This is a friendly and legible way of organizing the different types of information of each entity.

We start by the following definitions:

> k is the index of a terminal belonging to the set of the K terminals able to perform a handover, $k \in K$ and $\#k = K$;
> M_0 is the set of all possible PoAs, $M_1^{(k)}$ the set of all detected PoAs by the k terminal;
> $M^{(k)}$ is the set of PoAs that are allowed to the k terminal, $\#M^{(k)}=M_k$;
> W is the number of the properties of a PoA that will enter in the ranking process;
> $F^{(k)}$ is the set of all running flows of terminal k, being $\# F^{(k)} = N_k$ the number of running flows;
> *Flow map* allows mapping each of the N_k flows of a terminal to one PoA out of the M_k possible, $FM^{(k)}: F^{(k)} \rightarrow M^{(k)}$.

In order to model the three basic and independent entities in the architecture scheme (PoAs, users and flow maps), a specific matrix was define for each. The PoA profiles cover all the properties and context information about each PoA specifically. User profile relies on user/terminal preferences and on non real-time activity of the user, being totally independent of the PoAs properties. Flow maps are related with user's flows and with the resources available, being a kind of bridge between the information of the PoA and the user personal preferences and status.

PoA Profiles. Regarding PoA profiles, they are defined in this form $AP=(AP_{ij})_{MxW}$. This matrix keeps the PoAs properties and can be easily changed according to different criteria or preferences relevant in the mobility management decision. To keep the scheme architecture independent, we did not specify a method to set the mapping between numerical values and properties. However, a solution addressed by A. Iera et al. [5] is used. It is a simple analysis of each property setting an empirical numerical value to the criteria or being this value the result of a cost function.

The AP matrix is built based on all the specific properties of each PoA: taking this into account, its structure may be presented in three types of properties:

$$AP=(AP^{(user)}|AP^{(static)}|AP^{(real-time)})_{MxW}$$

The first substructure is set by information proceeding from the user, such as its preferences for the PoA. The *static* part refers to the properties of the PoA that, first of all, are static and independent of context, users, or time. The third part is built regarding the information that comes from the network, like the current resource status of the PoA.

Table 1. Possible PoA properties

Access Technology	User Preferences (user)	Monetary Cost (static)	Handover Effort (static)	Reliability (static)	Bandwidth Allocation (real-time)
UMTS	80	50	75	90	50
WiMAX	100	30	75	80	50
WLAN	70	80	100	40	50

An example of PoA properties and their empirical values is presented in [9] (these values are the ones of the AP matrix). In this matrix, the values closer to 100 are the best ones in that specific criterion. Bandwidth allocation can be a good example of a real-time property of a PoA, since it is dynamic and a result of a simple cost function, where the more occupied is a PoA, lower will be this value.

Flow Maps. The flow maps map the distribution of the different flows that belong to the same user through the available and allowed PoAs. Its mathematical model definition is:

$$FM^{(k,l)} = (FM_{ij})^{(k,l)}_{N \times M} \text{ such that } FM_{ij}^{(k,l)} \in \{0,1\}, \forall i, j$$

The l index defines a specific flow map for a given terminal k. Since we defined flow as the minimum indivisible unit of resources, $\sum_{j=1}^{M} FM_{ij}^{(k,l)} = 1, \ i = 1, ..., N$

User Profile. The user profile is based on properties and information independent of the context and real-time activity of the network. In order to have the proper interaction between the PoAs and the users, the user profile matrix must be modeled concerning the PoA properties:

$$UP^{(k)} = (UP_{ij})^{(k)}_{W \times W} \text{ such that } \left(UP_{ij}\right)^{(k)} = 0 \text{ if } i \neq j.$$

The UP is thus a diagonal matrix whose elements are weights that measure the importance given by the user k to the respective PoA criterion. It is now possible to shape qualitatively and quantitatively users using various combinations of different weights for each of the properties of the PoAs. Making a simple example by following this logic, different user profiles may exist such as *business man*, *gamer* and *groupie*. As is understandable and common sense, different users have different needs, and these requirements may be quantitatively weighted by the values in the UP matrix. Given this, a weight distribution like the present in [9] is adequate to model the type of users and their requirements.

Table 2. Possible Weight Distribution for Different User Profiles

User Profile	User Preferences	Monetary Cost	Handover Effort	Mobility Prediction	Bandwidth Allocation
Business Man	0.5	1.0	1.5	1.5	1.0
Gamer	1.5	0.5	1.5	1.5	1.0
Groupie	0.5	1.5	0.5	0.5	0.5

As is readily apparent, all these properties and values are easily configured and modified according to the criteria followed by the architecture designer, as planned in the design guidelines and requirements for the network selection scheme.

3.3 Network Selection Scheme

In this section we present the scheme for network selection – see Fig. 1. It has four main phases: (i) trigger management and processing, (ii) classification and prioritization, (iii) calculation of a set of ranked connectivity plans (try for a single terminal; if not possible, for a group of terminals) and (iv) handover initiation.

Trigger Management. Although a complete taxonomy of event sources is out of the scope for this work, one can think of three types of events: real-time events, scheduled and periodic. The first type is represented by the top-left corner of the diagram. Since complete information regarding the current communication state is split between the network and the terminal, both parties need to exchange perspectives. The diagram of Fig. 1 shows the network contacting the terminal (dashed arrow) and the terminal reporting information. It is also supposed to illustrate the situation where a change in context (wide sense) occurred. These changes can be detected by the network or the terminal so that any party can start a handover process. The second and third types of events (top-middle and top-right) are either periodic actions (such as denying access to some technology at a certain hour of the day) or scheduled actions.

Classification and Prioritization. Once a trigger is generated, it is classified and inserted in a common queue for processing. This design choice allows trigger differentiation, both in terms of trigger type and in terms of trigger owner. Regarding trigger type, we use two queues: urgent situations (e.g., terminal loosing signal) go to a Real-Time (RT) queue; non-urgent situations go to a queue named non-Real-Time (nRT). The main difference is the time needed to serve the request (e.g., 200ms for RT). The queue may also support user differentiation. This is useful in case of resource constraints: when determining flow maps, premium users should get their top choices.

Flow Maps Calculation. We enter now the stage where the algorithm calculates flow maps. There are two branches: local optimizations and global optimizations. Whereas the first concerns a single user, the second concerns the full optimization of a sufficiently large area and number of users. The distinction made is due to a simple reason: the (expected) heavy processing of the scheme. It is expected that, if not designed correctly, already single-user optimizations can be computationally heavy. We will describe the single-user branch (global branch is conceptually similar).

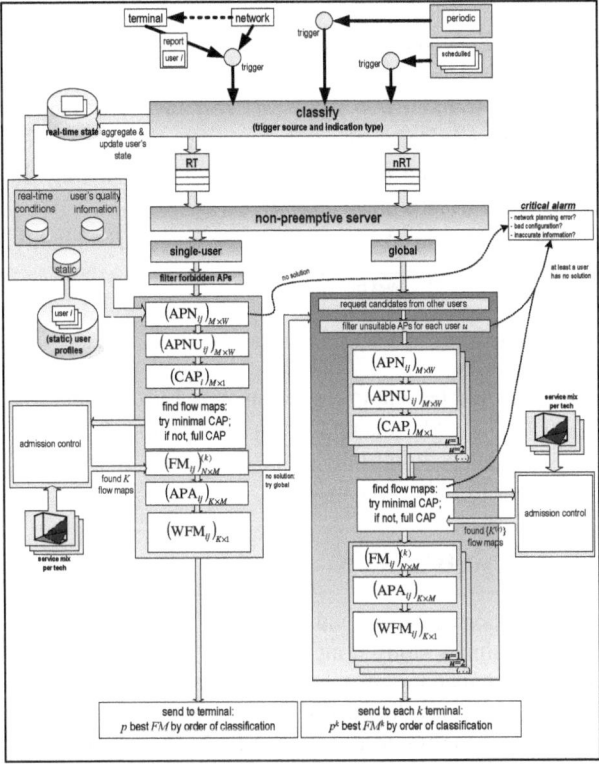

Fig. 1. Network Selection Scheme

This stage consists of algebraic manipulations of the matrices we have previously described and consists on the following steps:

1. Obtain a normalized version of AP to generate APN.
2. Find to which group the user belongs, retrieve the group profile and ate $APNU$.
3. Generate CAP using $(CAP_i)_{M \times 1} = \sum_{j=1}^{W} APNU_{ij}$: this is the overall cost of each PoA, already normalized and personalized.
4. Using the network's database of resources (must be updated), find the best S flow maps as described previously, according to the service mix deployed in the network. This scheme supports non-exhaustive algorithms: in order to reduce complexity, some PoAs may be concealed (which we called "minimal CAP"). This can be realistic in case the network is known to be always lightly loaded; if not, all solutions should be found to avoid running out of options.
5. (optional). In case no flow maps were found, generate alarm to network administrator and try a global optimization. This is conceptually similar to the case of a single user with the difference that now all flows from all users considered must be tentatively allocated to the reported PoAs (with the special constraint of not all users will see the same PoAs). The selection of terminals that belong

to the group can follow many strategies; perhaps the most obvious is by geo-graphical locatión and currently in reach of a set of PoAs known.

6. Suppose we have found S flow maps. Generate matrix "PoA Allocation" for each flow map found: $(APA_{ij})_{S\times M} = \sum_{m=1}^{N}(FM_{mj})^{(i)}$. This matrix determines how much used a certain PoA is for each flow map.

7. Determine the ranking of each flow map by generating the matrix WFM ("weights of flow maps") using $(WFM_i)_{S\times 1} = (APA_{ij})_{S\times M} \times (CAP_{ij})_{M\times 1}$.

8. Find in WFM the top p best ranked flow maps. The quality value of a flow maps is defined by $Q_i = \frac{WFM_i}{\max(WFM_i)}$.

3.4 Mobility Initiation

The best p flow maps should be sent to the terminal(s). Each terminal should receive the set of flow maps ordered by rank. It is up to each terminal to select a flow map according to undisclosed or local policies. It must, however, choose one from the set of offered ones; otherwise, all calculations must be done again.

4 Scenario and Results

This section presents a performance evaluation of the network selection scheme implemented. It also contains a study concerning the different parameters that can be configured in order to enhance the response of the global architecture.

A generic topology was developed for ns-2 2.31. This scenario is adjustable depending on the number of mobile terminals and PoAs, which are inputs of the topology file. The topology is based on a very simple wired-cum-wireless scenario. The topology dimension is calculated knowing the number of nodes in the network: distance of 700m between PoAs to avoid collisions and in order to emulate a multi-access technology scenario where the technologies do not interfere with each other. Links are defined to connect the fixed nodes, configured with 100Mb/s of bandwidth and a delay of 2ms. The wireless links are of 1Mb/s. For each mobile node it is created a User Datagram Protocol agent (UDP) and a Constant Bit Rate traffic (CBR) generation agent, transport and application respectively. Regarding CBR traffic, it is defined a rate of 100kb/s for every terminal, and a packet size of 1000 bytes, with each terminal generating/requesting traffic. Every presented value is a mean of 10 simulation runs and, whenever possible without compromising the correct visualization of the graphics, we provide the 90% confidence intervals.

4.1 Load Balancing

One of the real time properties of the PoAs is the resources availability (Bandwidth Allocation) at each moment in a specific access point. In the scheme implemented, this property is also considered, since it is expected that its utilization improves the performance of the architecture.

Introducing the maximum weight (1.5) for load balancing in the corresponding field of the user profile matrix, the global performance is the one shown in Fig. 2. As the number of PoAs increase, better performance is achieved, since there is a wider range of possible accesses and more available resources (with no delay for 10 PoAs).

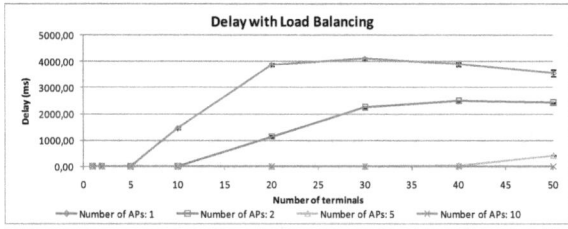

Fig. 2. Mean delay of scenarios with load balancing

4.2 Resource Management

The wireless channel in ns-2 2.31 is modeled to provide a maximum transfer rate of 1Mb/s, although in a real scenario this rate cannot be achieved without downgrading the quality of service provided. To evaluate this mechanism, we considered bandwidth thresholds for admission control ranging from 700kb/s to 1000kb/s.

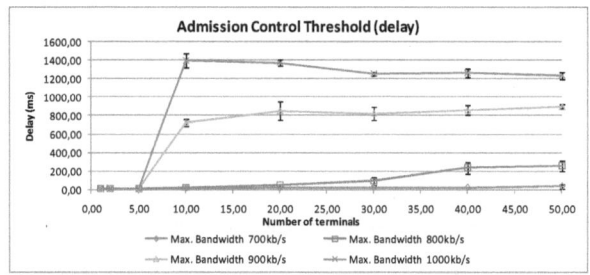

Fig. 3. Admission control thresholds comparison for delay

Fig. 3 depicts the packets delay achieved for different bandwidth thresholds that a PoA may allocate (with 5 PoAs). As expected, there is a clear tradeoff between traffic in the network and the quality of service provided. In the curves corresponding to 900kb/s and 1000kb/s bandwidth threshold, the value of the delay gets significant and stabilizes due to admission control (Fig. 4). As shown, the optimization algorithm filters the PoAs totally occupied, forbidding the terminals to connect to them even if they are the preferred ones. The number of blocked flows starts to increase as soon as the resources are all occupied in all PoAs.

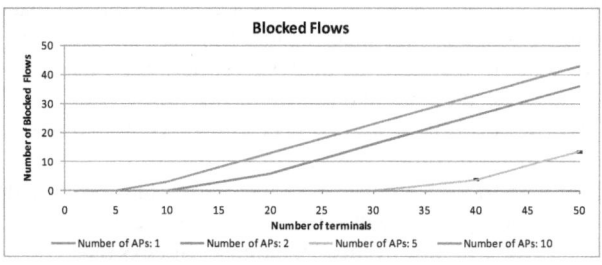

Fig. 4. Blocked flows with admission control

4.3 Triggers

As explained in section 3, the triggers are one main part of the optimization process, since they are the ones that initiate local or global optimizations. The decision on which optimizations should be performed may be configured through different criteria, besides the usual user requests that are considered a trigger.

To evaluate the effect of using triggers, simulations were performed based on the variation of the delay threshold used to trigger optimization. The scenarios tested were based on a threshold of maximum admissible bandwidth for admission control of 800kb/s, to be able to achieve significant delays and losses in order to trigger the optimization mechanism. These tests were performed using periodic QoS reports from the correspondent nodes at every second.

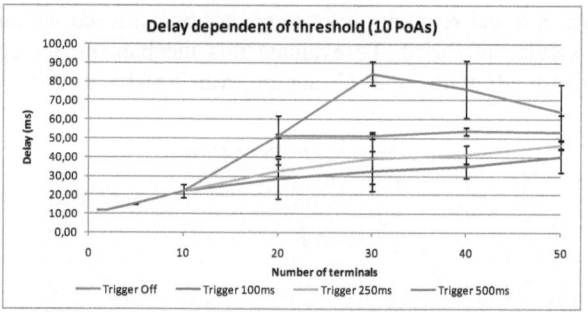

Fig. 5. Delay dependent of trigger thresholds

From the results obtained in Fig.5 in scenarios with 10 PoAs, it is possible to observe the improvements obtained with the utilization of QoS triggers. For the maximum number of terminals in each scenario, the network is never saturated, existing always available candidates for each terminal. As expected, for all scenarios, as the value of the delay trigger threshold decreases, the overall delay of the network also decreases (and also losses, not shown here). As depicted in Fig.5, the differences between the curves are significant. One interesting result (not shown here due to space limitations) is the non-significant influence in the overhead of the optimization process.

As a consequence of several triggers, the number of handovers increases degrading the loss ratio metric, as can be seen in Fig. 6. We observe that, for thresholds values of 100ms and 250ms, the number of triggers is higher; as a consequence, for these values the loss ratio will also be worse. This is a clear negative impact of performing many optimizations, although the values of loss ratio are not significantly high (maximum of 3%). In order to facilitate the analysis of the figure, the confidence intervals are not present, however they are about 50% of the mean value. Despite not shown, the impact in the overhead of using different thresholds for triggers is insignificant compared with the scenario without triggers off.

Directly related with triggers are the QoS reports from the correspondent node. These reports are sent periodically and are also responsible for the number of triggers in place during a simulation. We will now evaluate the impact of QoS reports rate in the network in scenarios with 10 PoAs available and with a delay threshold value of 250ms.

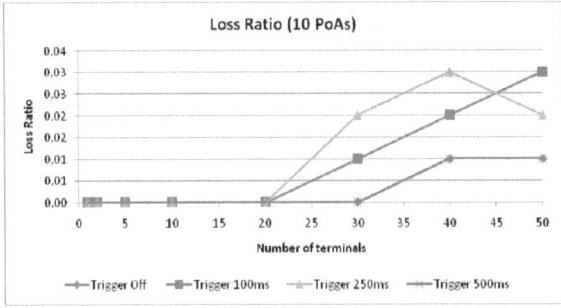

Fig. 6. Loss Ratio dependent of trigger thresholds

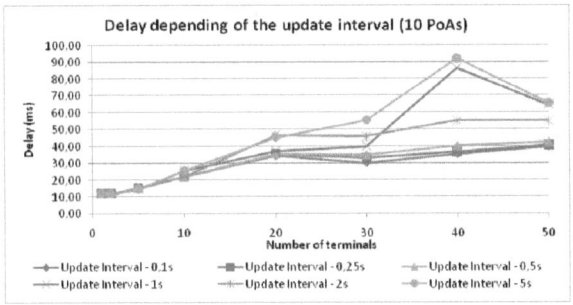

Fig. 7. Delay depending of QoS reports rate

As it is possible to conclude from the results in Fig. 7 there is an impact of the QoS reports rate in the network. However, it is not as evident as in the threshold case. For instance, establishing reports at every 0.1s is clearly better than configuring reports to each 5s, as expected. However, the difference between rates of 0.1sec, 0.25s and 0.5s is minimal. In this case, as opposed to the previous one, the difference in overhead is significant, because the reports can introduce large extra information in the network.

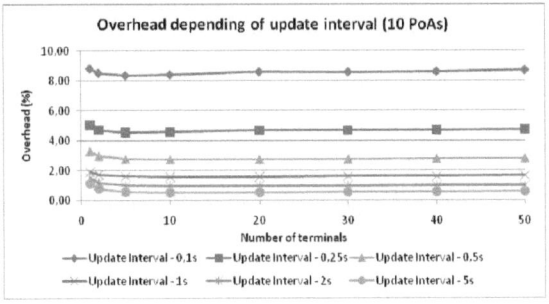

Fig. 8. Overhead depending of QoS reports rate

Regarding the overhead impact of this parameter, although there are clear benefits on using reports at every 0.1s, this causes a high overhead in the network (9%), which is clearly not a good solution, since it is consuming many resources. Once again, a tradeoff has to be made in order to achieve better delays but without overloading the network. Through the obtained results, a fair solution would be to configure the QoS reports with values around 0.5s or 1s. For the same reason that in Fig. 6, Fig. 7 and Fig. 8 do not show the confidence intervals, although their order of magnitude usually range from 10% to 20% of the mean.

4.4 User Preferences and Profile

In order to study the impact of the user preferences and profiles in the optimized deci-sions, a new metric was considered. It is defined as the ratio between the handovers performed to preferred PoAs and the total number of performed handovers.

The following results, depicted in Fig. 9, introduce a new parameter to the simula-tion which is the user profile, considering also different preferences of the terminal by each PoA. It is referred in the first case, a), that the results are the same for business and groupie profiles because the weight corresponding to the *user preferences* proper-ty given in their *user profile* matrices (UP) are the same. Both situations describe the impact of the load balancing weight in the preferred handovers ratio. For a null weight, the results are equal for both profiles, since the remaining parts of the APN matrix stay constant; the unique parameter that changes is the user preferences, irres-pectively of the weight given in the UP matrix (0.5 or 1.5) of each profile, because it will immediately determine the ranked list in function of this parameter.

However, as the load balancing weight increases, the ratio decreases, but in the gamer profile case, b), this decline is not as pronounced as in the business or groupie profile. This situation occurs due to the difference of weights given in the UP matrices and that influence the final ranked list of flow maps. As was previously mentioned, our mechanism enables the presence of this type of criteria in a seamless way.

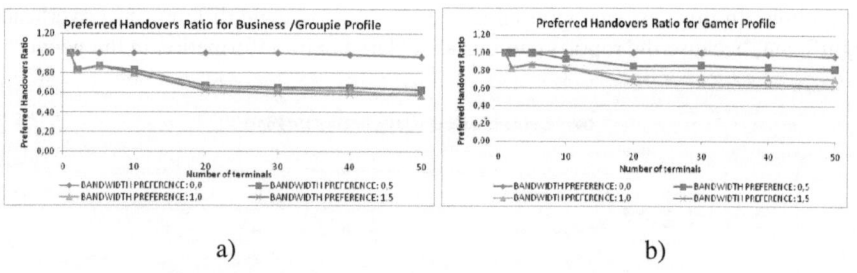

a)   b)

Fig. 9. Preferred handovers ratio for different user profiles

4.5 Global Optimization

Using an approach where the local optimizations just concern with resources availa-bility and terminal preferences, global optimizations can be used to re-organize the network. This process takes into account not only terminal's priority but also the state of current network resources through the load balancing feature. The results of this approach are present in Fig. 10. The scenarios were evaluated using a maximum bandwidth allocation per PoA of 800kb/s. It was also considered that all terminals have the same preferences and profile in order to be more evident the impact of global optimizations. The "After Global Optimization" situation occurs after a unique global optimization is performed after all flows are distributed. The periodic optimizations are scheduled to be executed in intervals of 5s after the simulation starts.

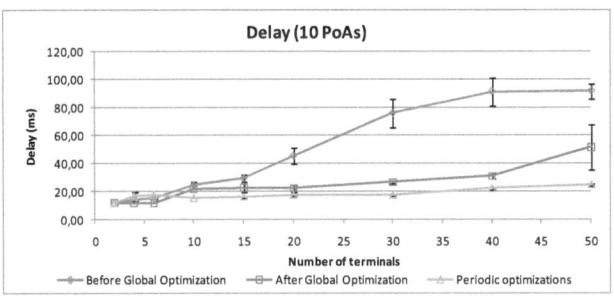

Fig. 10. Impact of Global Optimizations in scenarios with 10 PoAs

We observe in Fig. 10 that the benefits of performing global optimizations with load balancing are higher as the number of terminals increase. The curve corresponding to the mean delay before the optimizations, as expected, has higher values as the number of terminals increase. However, for scenarios with a low number of terminals, it tends to be nearer the other curves. Comparing the difference in the results between a unique optimization and periodic optimizations, the difference is not very sharp, with better results for the curve of periodic optimizations. However, as a global optimization always involves many handovers and re-allocations of flows, this solution may not be always the better for the user. This depends on many factors, and one of them is the efficiency and seamlessness of the mobility mechanism in place.

5 Conclusions

This paper presented an implementation and evaluation of the network selection algorithm developed to optimize the network performance in multi-technology and multihoming environments, considering the future paradigm of ABC in next heterogeneous networks. This optimization mechanism is able to process a decision based on different types of criteria, context, resources availability, QoS state, user profile and preferences, through a matrices formalism and a sequential process of algebraic manipulations to provide a ranked list with the best maps of flow's distribution through the available and allowed access technologies.

Through the performed experiments, we have shown that this mechanism is able to improve the overall network performance, being able to integrate several criteria in the optimization process. We also showed that a tradeoff needs to exist between the frequency of the triggers running the algorithm, in order to obtain good performance without increased overhead. As future work, we plan to integrate different types of networks in this mechanism, such as mesh and moving networks, and different types of services, such as multicast.

References

[1] Gustaffsson, E., Jonsson, A.: Always Best Connected. IEEE Wireless Communications 10(1), 49–55 (2003)
[2] O'Droma, M., et al.: Always Best Connected Enabled 4G Wireless World. IST Mobile Communications Summit 2003 (2003)
[3] Jesus, V., Sargento, S., Aguiar, R.L.: Any-Constraint Personalized Network Selection. In: IEEE Intl. Symposium on Personal, Indoor and Mobile Radio Communications, Cannes (accepted for publication) (September 2008)
[4] Prehofer, C., et al.: A framework for context-aware handover decisions. In: IEEE Intl. Symposium on Personal, Indoor and Mobile Radio Comm., Beijing (September 2003)
[5] Iera, et al.: An Access Network Selection Algorithm Dynamically Adapted to user Needs and Preferences. In: Proceedings of IEEE Intl. Symposium on Personal, Indoor and Mobile Radio Communications, Helsinki (September 2006)
[6] Furuskär, A., Zander, J.: Multiservice Allocation for Multiaccess Wireless Systems. IEEE Trans. on Wireless Communications 4(1) (January 2005)
[7] Gazis, V., Alonistioti, N., Merakos, L.: Toward a Generic "Always Best Connected" Capability in Integrated WLAN/UMTS Cellular Mobile Networks (and Beyond). IEEE Wireless Comm. 12(3), 20–28 (2005)
[8] Xing, B., Venkatasubramanian, N.: Multi-Constraint Dynamic Access Selection in Always Best Connected Networks. In: IEEE MobiQuitous 2005, CA (July 2005)
[9] Jesus, V., et al.: Mobility with QoS Support for Multi-Interface Terminals: Combined User and Network Approach. In: IEEE Symposium on Computers and Communications, Aveiro, Portugal (July 2007)
[10] Pahlavan, K., Krishnamurthy, P., Hatami, A., Ylianttila, M., Makela, J.P., Pichna, R., Vallstron, J.: Handoff in Hybrid Mobile Data Networks, Personal Communications, vol. 7, pp. 34–47 (2000)
[11] Stevens-Navarro, E., Wong, V.W.S.: Comparison between Vertical Handoff Decision Algorithms for Heterogeneous Wireless Networks. In: Proc. of IEEE Vehicular Technology Conference (VTC-Spring 2006), Australia (May 2006)
[12] McNair, J., Zhu, F.: Vertical handoffs in fourth-generation multinetwork environments. IEEE Wireless Communications 11(3), 8–15 (2004)
[13] Chen, W.-T., Shu, Y.-Y.: Active application oriented vertical handoff in next-generation wireless networks. In: Proc. of IEEE Wireless Communications and Networking Conference, New Orleans, USA (March 2005)
[14] Song, Q., Jamalipour, A.: Network Selection in an Integrated Wireless LAN and UMTS Environment using Mathematical Modeling and Computing Techniques. IEEE Wireless Communication 12(3), 42–48 (2005)

Performance of Optical Ring Architectures with Variable-Size Packets: In-Line Buffers vs Semi-synchronous and Asynchronous Transparent MAC Protocols[*]

Thierry Eido[1], Tuan Dung Nguyen[1], Tülin Atmaca[1], and Dominique Chiaroni[2]

[1] TELECOM & Management SudParis, RST Dept., 9 rue Charles Fourier,
Evry, 91011 France
[2] Alcatel-Lucent Bell Labs, Route de Villejust, 91620 Nozay, France
{Thierry.Eido,Dung.Nguyen,Tulin.Atmaca}@it-sudParis.eu,
Dominique.Chiaroni@alcatel-lucent.fr

Abstract. The rapid growth in client application demands, in terms of bandwidth and Quality of Service (QoS), has motivated the deployment of the optical technology at Metro Access and Metro Core Networks. A new Ring-based Optical Network Architecture is being designed by Alcatel-Lucent with a view to offer a cost effective solution that provides also some added values in terms of reliability and performance. In this paper, we propose a packet size-based scheduling algorithm and a priority-based Packet Erasing Mechanism (PEM). These two mechanisms improve the performance (in terms of packet loss ratios and mean access delays) of the proposed optical ring architecture. Finally, through simulation works we present a performance analysis and comparison of three different optical ring architectures in competition. Obtained results provide interesting conclusions on the behavior and performance tendencies for each one of the studied architectures.

Keywords: MAC Protocols, Optical Metro Access Ring Networks, Optical Packet Switching (OPS), Performance Analysis, Variable-Size Packet Format.

1 Introduction

Optical Packet Switching (OPS) technologies are among the most promising solutions for Next Generation Network architectures. In recent years, we have witnessed a rapid growth of client application demands in terms of bandwidth and Quality of Service (QoS). The deployment of the optical technology at Metro Access and Metro Core Networks is becoming more and more rather a necessity than an option.

Ring-based topologies can be well adapted for Metro Access and Metro Core Networks since a ring is simpler to operate and administrate than a complex mesh network. Packet ring-based networks were pioneered by the Cambridge Ring [1],

---

[*] This work was partially supported by the French ANR / ECOFRAME and by the European EuroNF research projects.

R. Valadas and P. Salvador (Eds.): FITraMEn 2008, LNCS 5464, pp. 169–184, 2009.
© Springer-Verlag Berlin Heidelberg 2009

followed by other important network architectures such as Meta-Ring [2], Token Ring [3], FDDI [4], ATMR [5], CRMA-II [6] and RPR [7]. Those networks are mainly based on electronic routers which provide satisfying network performances but with relatively high cost.

However, Alcatel-Lucent is targeting a metro access network capable to evolve towards a metro core plate-form [8], [9]. The big challenge is to find a new technological approach that provides a solution at lower cost compared to a pure electronic solution, but has also some added values in terms of reliability and performance.

This paper provides a performance comparison of three optical ring architectures that transport variable-size packets (Fixed-size packet architecture is studied in one of our recent works [13]). These architectures are differentiated by the way transit packets at intermediate nodes are handled. The first studied architecture is a pure electronic reference model which operates in a way which is similar to RPR (Resilient Packet Ring). All transit traffic is demodulated at each intermediate node, converted to the electronic domain and stored inside in-line electronic buffers. Packets waiting in the in-line buffers enter in competition with local packets, according to a priority scheme, in order to be forwarded to the next node on the ring.

However, in both second and third studied models transit packets bypass intermediate nodes without being demodulated or processed locally. This feature offers considerable gain in terms of network cost since each node should be equipped only with enough capacity for the transport of its own traffic. The difference between these two later models lays on the fact that one uses an asynchronous transmission mode for insertion of new packets on the ring and the second uses a semi-synchronous one (proposed architecture described in section 2.2).

Furthermore, we propose two mechanisms that improve the performance of the proposed optical ring architecture. Because of the random access to the ring and size-variability of transported packets, free spaces on the ring (available for transmission) are also of variable size. Thus, a packet size-based scheduling mechanism is proposed with a view to avoid situations where head of line packets of large size block the transmission of smaller packets waiting behind. The second proposed mechanism is a threshold-controlled Packet Erasing Mechanism (so called PEM). It consists of erasing, at intermediate nodes, some best effort packets in transit, to the favor of insertion of local premium traffic which performance can be degraded otherwise.

Finally we provide, through simulation works, a performance comparison (in terms of packet loss ratios, access and transit delays) of the studied optical ring architectures. We investigate also the performance gain obtained by the proposed scheduling and PEM mechanisms.

The rest of this paper is organized as follows. Section 2 presents the studied network architectures, including the reference electronic model based on RPR, the Alcatel-Lucent optoelectronic model and the asynchronous passive optical ring model (DBORN). Section 3 describes both our packet priority-based erasing mechanism (PEM) and our packet size-based scheduling algorithm. Sections 4 and 5 are dedicated to the description of performance parameters a set of simulation results comparing the performance of the aforementioned architectures. Finally, section 6 concludes this paper and provides some perspectives on future work.

2   Studied Network Architectures

2.1   Reference Electronic Model

Figure 1 describes the first reference model with pure electronic nodes. This model is based on RPR [7].

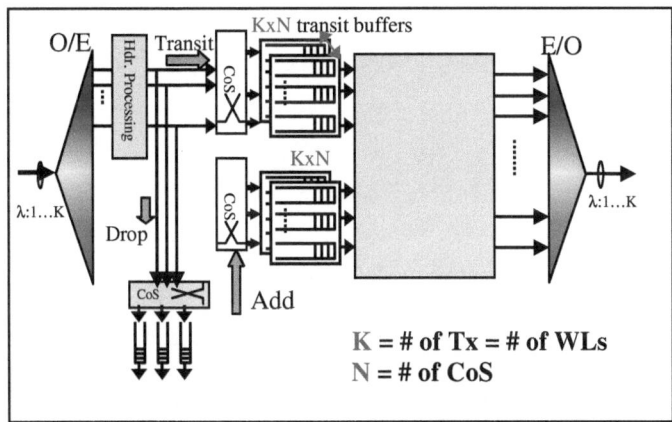

Fig. 1. Logical model of the reference electronic model

In this model, each node has access to all wavelengths on the ring (K wavelengths). The network supports a number of N classes of service. All transit optical packets (coming from the previous node on the ring and not destined to the current node) are converted into electronic packets and stored in transit buffers ($K*N$ transit buffers) before being converted again to optical signals and transmitted on the ring. The "Add traffic" is formed by client packets received at the node aggregation points and stored in local buffers until their transmission on the ring begins. A priority policy is applied to transit and add traffic according the following rules:

- For the same type of traffic (add or transit), traffic of class i has a priority over traffic of class j, for all $j > i$.
- Transit traffic of class i has a priority over add traffic of class j, for all $j \geq i$.
- Add traffic of class i has a priority over transit traffic of class k, for all $i < k$.
- Overall, the electronic switch serves packets from all buffers with a simple head-of-the-line (HOL) priority scheme in a First-In-First-Out (FIFO) order.

2.2   Novel Semi-synchronous Active Optoelectronic Model

Figure 2 shows the logical model of a ring node in the studied semi-synchronous optoelectronic architecture, including the way each node is connected to the ring and the associated medium access protocol.

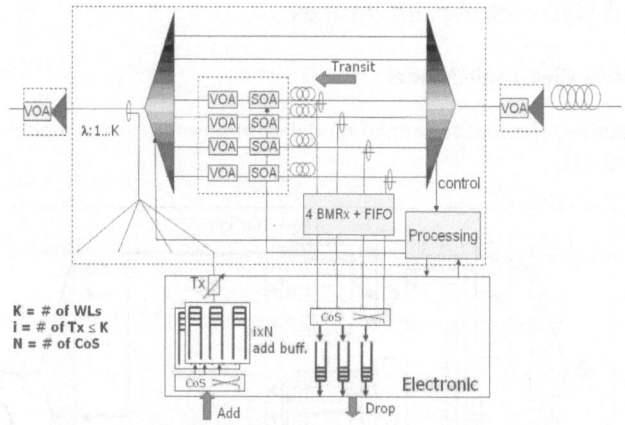

Fig. 2. Logical model for the optoelectronic node

The structure of the optical part of ring nodes mainly consists of accordable optical transmitters and burst mode receivers (i.e., accordable transmitters (Tx) and burst mode receivers (BMRx) working in an asynchronous mode). The accordable transmitters offer to the system a modular architecture. In fact, the node traffic volume is usually much smaller than the network transmission capacity. Therefore, each node is equipped with a number of accordable transmitters, which is smaller than the number of wavelengths in the ring. The node capacity can be increased by inserting modular optoelectronic cards, each one has an accordable transmitter and N local buffers (N is the number of classes of services supported by the network).

The modular aspect in the optoelectronic model reduces the network cost compared to the electronic reference model, where each node is designed to support a traffic volume which is equal to the ring capacity. Moreover, each node in the electronic reference model has to demodulate and re-modulate all the transiting traffic, thus increasing the processing time and power consumption.

At the optical level, ring nodes use an optical de-multiplexer to separate the optical spectrum into several wavelengths that can be used by the node for packet insertion/reception. A fiber delay line (FDL) creates on the transit path a fixed delay between the moments when the control process is executed (packet header analyzing, etc.), and the Add/Drop function. The Add function is done by the local accordable Tx, while the Drop function is performed by the Semiconductor Optical Amplifiers (SOA) triggered when a signal is received from the control processing unit. Packets destined to the node are filtered and received on burst mode receivers and new local packets are inserted into detected voids by accordable transmitters. Note here that the implementation of add / drop functions in the same optical node is henceforth feasible thanks to the use of new devices such as the SOA components that can be controlled by the processing unit.

Since this ring architecture is designed for both metro access and metro core networks, signal amplification should be involved at intermediate points on the ring. In order to lower the cost of required optical devices, a constant level of signal should be maintained on the ring. This idea drives us to the introduction of the concept of "empty packets".

An empty packet can be defined as a modulated signal, which represents a piece of free bandwidth. An empty packet has the same format as data packets (header with packet size, etc.), however the priority field in an empty packet is set to a special value (for example 0) to indicate that the packet is not a data packet. In our optoelectronic model, we use empty packets in order to simplify the optical amplifying system.

Initially, the ring is filled with a continuous stream of empty packets (this operation is performed by an arbitrary node, which must be equipped with as many transmitters Tx as the number of wavelengths in the ring). Optical data packets are of variable-size. When a data packet is inserted to the ring, it replaces one or several empty packets. If the space of the last empty packet is not completely used by the inserted packet, a smaller empty packet is generated so that the signal stream will remain continuous.

When a packet is dropped at the transit line, the released space can be used for the insertion of local packets. If this space is not completely used for local insertions, the remaining space is filled by an empty packet. Note here that transmission of empty packets cannot be interrupted. Insertion of local packet stream can start only at the beginning of one or several replaced empty packets. This feature gives to this architecture an aspect of synchronization. Figure 3 shows the insertion process using empty packets.

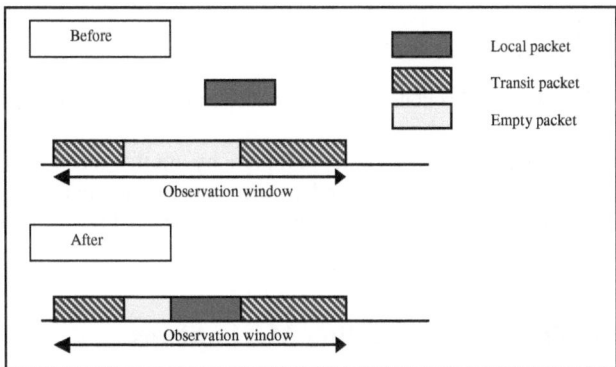

Fig. 3. Voids management for packet insertion

The studied optoelectronic solution has many advantages over existing solutions, such as RPR or NG-SDH/SONET:

- It relies on shared optical wavelengths, which contributes in reducing the number of transponders required in the network because ring nodes have transponders for their local traffic only (no need for O/E/O conversion for the transiting traffic).
- Tx and Rx are out of the transit line, making easier the node upgradeability, limiting the service interruptions and reducing protection cost (1:N vs. 1:1).

2.3 Asynchronous Passive Optoelectronic Model (DBORN)

The third architecture studied in this work is DBORN (Dual Bus Optical Ring Network). It consists in a passive version of the proposed optoelectronic model, in which the ring is divided into two different buses: *"Upstream Bus"* and *"Downstream Bus"*. The communication between ring nodes is centralized by a *"Hub"* node. The transmission mode is asynchronous. Ring nodes insert their traffic on the upstream bus, where packets are transmitted without interruptions, all along their way to the Hub node. The Hub node then delivers packets to their destination through the downstream *"reception"* bus.

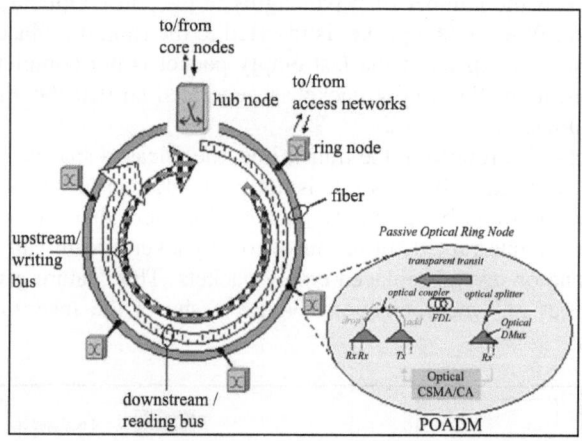

Fig. 4. Asynchronous Ring Network Architecture (DBORN)

DBORN architecture can be typically adopted by metro access networks, where client traffic is collected then transmitted to clients of other networks, through the centralized Hub node. A more detailed description of the DBORN architecture can be found in reference [12].

3 Packet Erasing and Scheduling Mechanisms

3.1 QoS Management and Packet Erasing

In the proposed optoelectronic architecture (section 2.2), each node has several add electronic buffers which can store the incoming client packets. Add local buffers are associated to the supported classes of service (one buffer for each class of service). When the transit line is free, these buffers are served according to a Head of Line (HOL) policy. As a general rule, transit traffic cannot be interrupted by the local traffic, except for the case of packet erasing mechanism, defined as follows. The erasing mechanism prevents higher priority packets from being lost or having excessive access delay. When this mechanism is enabled, Best Effort packets at the transit line can be dropped in order to insert higher priority packets waiting in the Add buffer. The activation of the erasing mechanism is controlled by a threshold S, which represents a critical filling level of higher priority buffers (i.e. the erasing mechanism is enabled only when the filling level of higher priority buffer is superior to S).

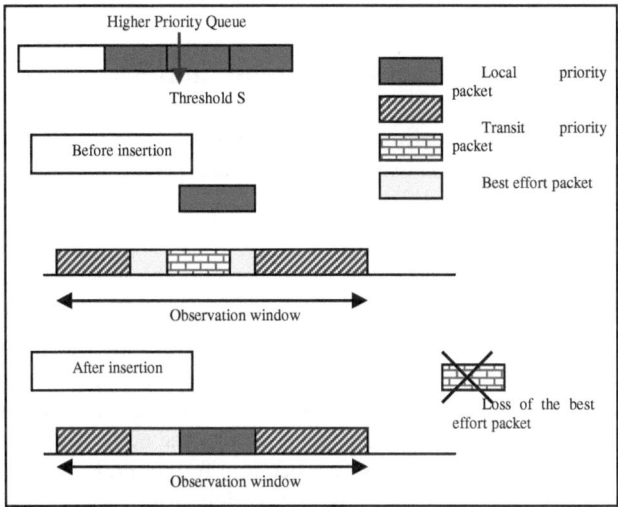

Fig. 5. Packet erasing mechanism

3.2 Size-Based Packet Scheduling

As we have explained in the previous sections, when a sufficient free space is available on the fiber, a packet waiting at the Add buffer is inserted into the ring according to a Head of Line (HOL) policy. However, this policy presents a blocking problem in the case of variable packet-size architecture. In fact, when the free space on the fiber is not big enough to transmit the first packet of the queue, no packet insertion is performed. Though, in the same queue, some packets may have a smaller size which fits into the available void on the fiber.

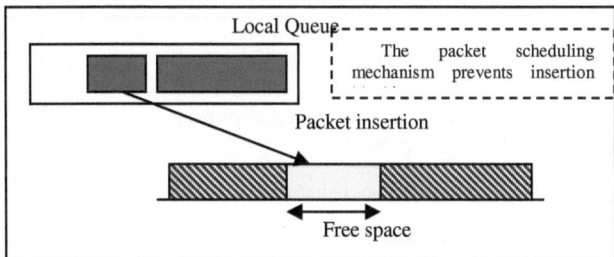

Fig. 6. Size-based packet scheduling mechanism

In order to improve the packet sending in our optoelectronic model, we propose a size-based packet scheduling mechanism. When activated, this mechanism allows the control unit to choose for insertion any packet waiting in the queue and not only the first one. In our experimentations, we limit this choice to the first k packets in each queue (after some simulation experiments, we chose to set k to 3 packets). This allows us to limit the additional processing time due to buffer scanning procedure. Figure 6 illustrates the principle of our packet scheduling mechanism.

4 Performance Parameters and Simulation Scenarios

This section consists in evaluating and comparing the performance of the above described network architectures. All simulations are performed using discrete event network simulation tool (ns 2.1b8 – [11]).

4.1 Performance Parameters

In order to evaluate the performance of the above described architectures, we use the following performance metrics.

Global Packet Loss Rate (Global PLR): The Global PLR at each node is defined as the ratio of lost client packets due to buffer overflow and due to erasing mechanism (for transit packets) to the total packets generated by the node. This loss is seen by the source node, even though loss events may occur at further nodes on the ring.

Average Access Delay: The Average access delay of an electronic packet at each node is defined as the average waiting time measured from the moment when the packet enters in the local buffer till its transmission begins.

4.2 Evaluation Scenarios

For the service differentiation, we consider the metro Ethernet Service Level Agreement (SLA) as in reference [11]. Four main optical classes of service are considered in order to accommodate eight client classes of service determined per CoS-ID, according to the IEEE 802.1 specifications.

Table 1. CoS Identification

CoS	Supported services	Performance
Premium CoS1-2	Real time telephony or video applications	Delay < 5ms Jitter < 1 ms Loss < 0.001%
Silver CoS3-4	Bursty mission critical data applications requiring low loss and delay (e.g. Storage)	Delay < 5ms Jitter = N/S Loss < 0.01%
Bronze CoS5-6	Bursty data applications requiring bandwidth assurances	Delay < 15ms Jitter = N/S Loss < 0.1%
Standard CoS7-8	Best effort service	Delay < 30ms Jitter = N/S Loss < 0.5%

The table 1 shows our reference classes of service (CoS) in the optical domain. The first column indicates, for each considered optical class of service, the ID of electronic classes which are going to be accommodated inside (CoS mapping scheme).

We simulate the network with 10 ring nodes transmitting on one wavelength of 10 Gbit/s. The offered network load is set to 0.75, and the traffic is generated uniformly by ring nodes. We assume that the propagation delay between adjacent nodes is 0.2 ms, which is equivalent to some 40 km (close to the real size of a metro network). We suppose that the optical header is composed of 16 bytes of preamble and 32 bytes of inter-packet gap.

For the premium traffic, we consider constant bit rate sources, with a packet size of 810 bytes. This assumption comes from the fact that today's premium service such as voice, video are transported mainly over SONET/SDH networks, which are constant bit rate sources. The non premium traffic is modeled by an aggregation of IPP (Interrupted Poisson Process) sources with different burstiness levels. Packets generated by these sources are of variable length. According to the Internet packet length statistics [14], sizes of these packets vary from 50 bytes to 1500 bytes.

Table 2. Traffic Hypothesis at Electronic Client Side

	CoS1-CoS2 Premium	CoS3-CoS4 Silver	CoS5-CoS6 Bronze	CoS7-CoS8 BE
%CoS	10.4% x 2	13.21% x 2	13.21% x 2	13.21% x2
Packet size (bytes)	810	50, 500 & 1500	50, 500 & 1500	50, 500 & 1500
Burstiness	CBR	IPP b=5	IPP b=5	IPP b=10
Burst Length	CBR	10	10	20
Local Buffer Size (Kbytes)	100	250	250	500

Local buffer sizes are respectively equal to 100 Kbytes for premium traffic classes, 250 Kbytes for silver and bronze traffic classes and 500 Kbytes for Best Effort traffic classes. Only the premium and silver classes of traffic can take benefit of the erasing mechanism. Erasing thresholds for these classes are respectively S1 and S2.

5   Simulation Results

5.1 1-N Scenario

In this Scenario, we consider 10 nodes communicating symmetrically through the ring. Each node generates 10% of the total traffic offered to the network (i.e., uniform traffic partition). The traffic matrix is a balanced full-meshed matrix. This means that each node sends equal amount of traffic to all other nodes in the ring. The offered network load is set to 0.75 (i.e., the traffic offered to the network is set in a way that each link will be loaded to 0.75 of its capacity). For the Alcatel-Lucent model analysis, we consider 5 study cases:

Case A: The PEM erasing mechanism is enabled with the following rule: a transit packet of CoS 7 or 8 is systematically destroyed for the insertion of a local packets of CoS 1, 2, 3 and 4 (No threshold control).
Case B: PEM mechanism is enabled with the following rule: a transit packet of CoS 7 or 8 is destroyed for the insertion of a local packet from CoS 1, 2, 3 and 4 according to a threshold: S1 = 800 bytes and S2 = 1500 bytes.
Case C: the same rules as in case C with different threshold values: S1 = 2000 bytes and S2 = 5000 bytes.
Case D: erasing mechanism is disabled.
Case E: In addition to the rules used in case C, we also use the size-based packet scheduling mechanism described earlier.

In this experimentation, because of the symmetric nature of the traffic matrix, performance results are similar for all ring nodes. Thus, we have limited the results analysis to one single node.

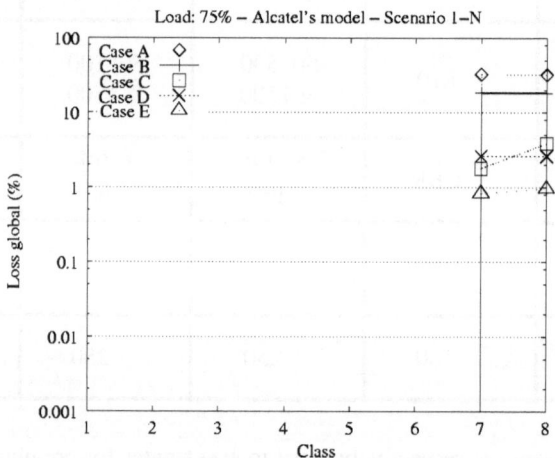

Fig. 7. Global PLR at each node for optoelectronic model

Figure 7 shows the global PLR of the traffic generated by a single ring node for each client class of service. Note that in this simulation case, there is no packet loss observed for the reference electronic model (described in section 2.1). As for the optoelectronic model, losses are observed only for the CoS 7 and 8 for all simulation cases (thus, QoS for premium services is preserved). This has two reasons, the first one is that the erasing mechanism causes loss at the transit lines only for best effort traffic (classes 7 and 8), and the second reason is that the CoS 7 and 8 have the lowest priority among local Add traffics.

Simulation results show that a small erasing threshold causes excessive PLR at the transit lines, while a higher threshold provides a lower PLR. On the other hand, Figure 8 shows that when the packet erasing mechanism is limited by a higher threshold, the access delay is globally increased for all classes of service. This is due to the fact that a higher erasing threshold reduces the probability of erasing transit packets for insertion of high priority local packets, thus increasing the waiting time of high priority local packets, hence that of lower priority packets.

Figures 7 and 8 show that the size-based packet scheduling mechanism, combined with a high erasing threshold value, provides both reasonable PLR and average access delays. However, the size-based packet scheduling mechanism may need additional processing time.

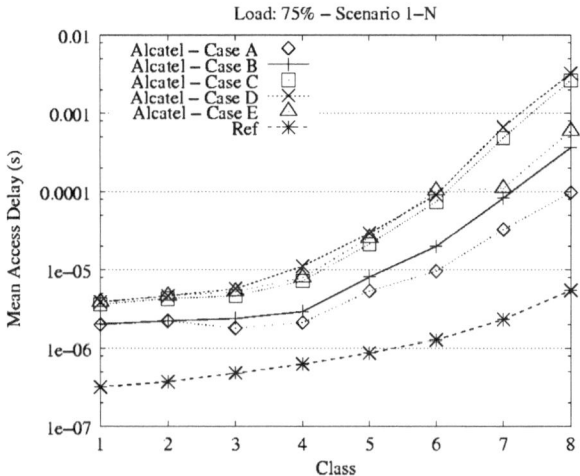

Fig. 8. Mean access delay for each CoS

Regarding the access delay results (Figure 8), we first notice that the proposed optoelectronic model offers higher access delay for all CoS packets compared to the reference electronic model. This is due to the fact that the reference architecture exploits better the available bandwidth thanks to the availability of electronic memory, thus avoiding the well-known problem of bandwidth fragmentation which typically occurs in optical ring networks with random access to the transmission medium. We understand by bandwidth fragmentation, the phenomenon where the amount of bandwidth which is available for use, in a wavelength becomes divided into many small pieces [6]. This is due to random packet insertions on a shared medium. Therefore, free spaces which are smaller then the optical packet size are out of use for traffic insertion.

Note that this problem of fairness and bandwidth fragmentation has been studied carefully in precedent strategic collaborations on DBORN architecture [12].

However, Figure 8 also shows that thanks to the packet erasing mechanism introduced in the Alcatel-Lucent architecture, the performance of priority services, namely CoS1 to CoS6, are guaranteed. For instance, with systematic erasing policy (case A), the mean access delays of CoS1 to CoS6 remain lower than 10 μs. Note that this property did not exist in passive node architecture such as DBORN, where the performance of premium services at downstream nodes is highly impacted by the increase in volume of best-effort traffic at upstream nodes.

5.2 Bus Scenario

In the following subsections, we will investigate performance analysis of the described architectures through a bus-based network scenario, which is typically the case for the access metro network. This means that 9 access nodes send traffic on the bus (one wavelength at 10Gbit/s) to one centralized node. The offered network load is set to 0.75, and the traffic is partitioned uniformly on the bus.

5.2.1 Comparison of the Proposed Model Optoelectronic Model to the Electronic Reference Model

For the optoelectronic model, we use the following values for erasing thresholds: S1 = 2000 bytes and S2 = 5000 bytes.

Figure 9 shows the global PLR for each CoS at each access node for the optoelectronic model. Note that no packet loss has been observed in case of electronic reference model. As for the optoelectronic model, we only observe losses of CoS 7 and 8 (best-effort traffic) mainly at the transit line (due to packet erasing mechanism) for nodes 0 to 7, and at the Add buffers for the last node on the bus (Node 8).

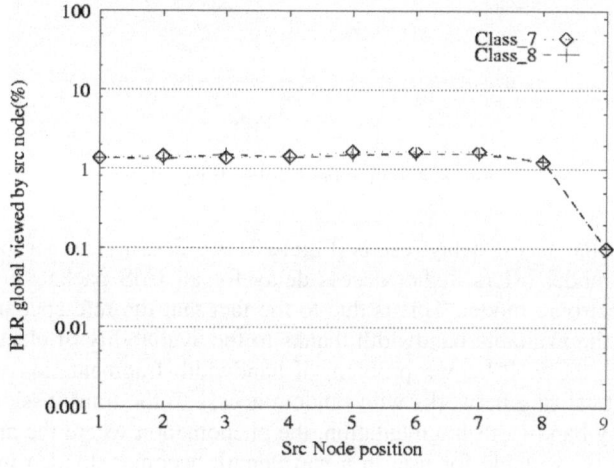

Fig. 9. Bus scenario – Global PLR at each node for Alcatel optoelectronic model

Figure 10 presents the mean access delay of each CoS at each node in Alcatel optoelectronic model and reference electronic model. Globally, the access delays of all CoS in both models increase as we move downstream on the bus. This is due to the well-known unfairness property in bus-based network, in which upstream nodes own the total bandwidth and may block the transmission of downstream nodes, leading to the increase of access delay at those nodes [12].

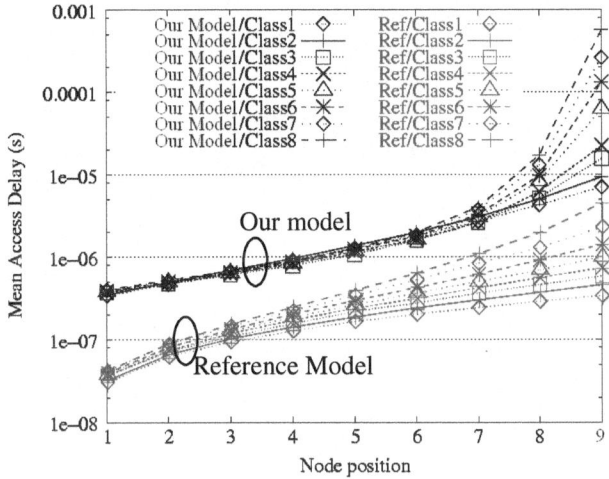

Fig. 10. Bus scenario – Mean access delay at each node for each CoS

For both models, we observe that thanks to the priority policy, the prioritized traf-fic (e.g., CoS 1 to 4) experience low delay even at the last node, while the lowest priority traffic (CoS7, 8) suffer from the lack of bandwidth and experience highest access delay. Moreover, for optoelectronic model, the CoS7 and CoS8 see their delay exploded at the last access node.

We observe that in all cases the access delay obtained in reference model is always lower than that obtained in optoelectronic model, since access nodes in the optoelec-tronic model suffer from the bandwidth fragmentation phenomenon as explained in previous section. In this experimentation, one may notice that for the first node on the bus, which owns the total network bandwidth, the access delay obtained in reference electronic model and in optoelectronic model must be the same. Nevertheless, in Figure 10, we see a difference in access delay at the first node. This is due to the fact that in the optoelectronic model, we use the empty packets for the MAC protocol. With this technique, a node with empty buffer must generate an empty packet for maintaining the signal synchronization. During the transmission of empty packets, any arriving client packet will be stored and wait for transmission until the end of this empty packet. Therefore this feature creates an additional delay to the mean access delay. This additional delay would not exist if the network used the void detection mechanism for the MAC protocol (as in DBORN). We will investigate this problem later in the next paragraphs.

5.2.2 Comparison between DBORN and Optoelectronic Model

As stated earlier, the technical difference between the use of empty packets and voids for MAC protocol may introduce differences in terms of performance between these architectures. This subsection investigates the performance comparison of the Alcatel optoelectronic model (with empty packets) and DBORN architecture (with void detection).

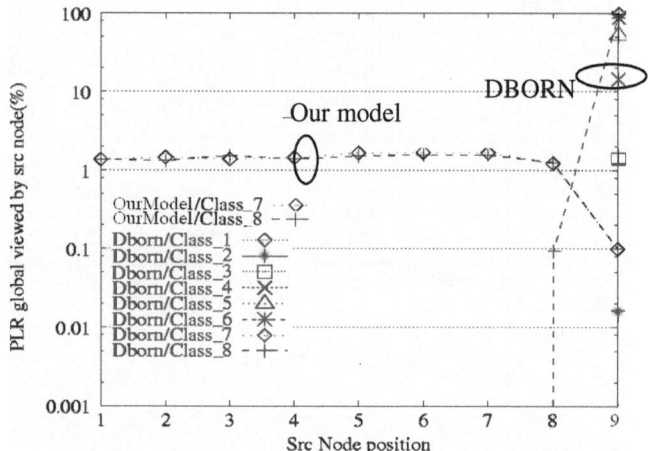

Fig. 11. DBORN vs. Optoelectronic model: Global PLR

Figure 11 plots the global PLR at each node for both Alcatel optoelectronic model and DBORN. As we can see, in Alcatel optoelectronic model, there are small losses only of best-effort packets but at each node. On the other hand, with DBORN architecture, we observe losses of packets only at nodes 7 and 8, with some small PLR for CoS2 and CoS3, and very high PLR for CoS4 to CoS8 traffics (from 10% to nearly 100% respectively). With DBORN, losses at node 7 and 8 are due to add buffers overflow. Moreover, the PLR in DBORN explodes for low priority traffic at the last bus node, while the PLR in Alcatel-Lucent new optoelectronic model only concerns best-effort traffic and remains less than some percents. Therefore, we can deduce that the problems of unfairness and bandwidth segmentation (well-known for this bus-based architecture) are lightened with the use of empty packets and packet erasing mechanism in Alcatel-Lucent optoelectronic model.

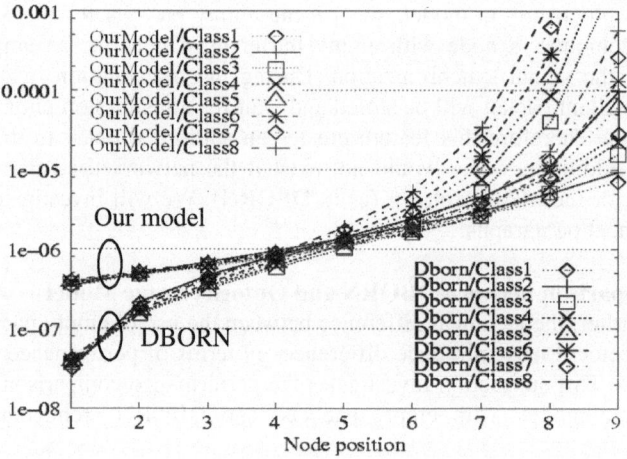

Fig. 12. DBORN vs. Optoelectronic model: Mean access delay at each node

Figure 12 presents the mean access delay at each access node for DBORN and optoelectronic models. Because of the intrinsic graduated loading property of bus-based network, the mean access delays in both models rapidly increase at the last nodes.

However we observe that, compared to the DBORN model, the optoelectronic model provides higher access delay at some upstream nodes but very lower access delay at last downstream nodes. Thus, the use of empty packets and packet erasing mechanism lightly balances the node access delays on the bus. A possible explanation for this behavior is that since the transmission of an empty packet cannot be interrupted by the arrival of an Add packet, it increases the access delay at some upstream nodes. At the same time, this empty packet may be useful for other downstream nodes to insert an Add packet. This forms a kind of resource preservation, by the nature of the MAC operation, hence lightly increasing the available bandwidth for downstream nodes.

6 Conclusion

In this paper, we provided a comparative performance study of the new optoelectronic node architecture proposed by Alcatel-Lucent. Our analysis was mainly based on simulation results comparing the later with a reference electronic node model (e.g. RPR standard) and a passive asynchronous ring model (e.g. DBORN).

Compared to the full electronic node approach, the new optoelectronic model provides significant network cost reduction, better reliability for the transit path, and better modularity for the node. We have proposed a QoS management mechanism: Threshold-based Packet Erasing Mechanism (PEM). It allows packets of higher priority to preempt transit best effort packets and take their place for being transmitted on the ring. The optimum threshold value would allow the erasing of best effort packets only when priority service performance index reach certain alarming levels.

We have also proposed a size-based packet scheduling mechanism applied to packets in the Add buffers in order to prevent bandwidth waste that occurs in variable packet length architectures, when the Head of the Line (HOL) service is used. This mechanism contributes in bringing back the network performance to a safer level, by providing a better use of bandwidth resources. However, the use of this mechanism may render more complex the node architecture and cause higher processing delay at access nodes. A trade-off between performance and complexity should made.

We have performed a comparison between the use of empty packets (in the new Alcatel-Lucent model) and the simple void detection in MAC protocol (e.g. in DBORN). Performance results show that the use of empty packets contributes to the balancing of upstream and downstream nodes performance (access delay and PLR). A possible explanation for this behavior is that since the transmission of an empty packet cannot be interrupted by the arrival of an Add packet, it increases the access delay at some upstream nodes. At the same time, this empty packet may be useful for other downstream nodes to insert an Add packet.

Note that from the performance perspective, the use of empty packet is preferable in case where the network supports variable length packet format. In case of fixed length packet format with semi-synchronous transmission mode, the packet transmission will be always synchronized at the beginning of (free) time slots, regardless of the definition of a time slot (a fixed void or a fixed empty packet).

However, since the study was focused on variable length packet format, the problem of bandwidth segmentation still arises in ring architecture. The fixed length packet format with semi-synchronous transmission may provide a solution for the bandwidth segmentation problem, but it may introduce another problem: the optical packet filling efficiency.

References

1. Needham, R.M., Herbert, A.J.: The Cambridge Distributed Computing System. Addison-Wesley, Reading (1982)
2. Cidon, I., Ofek, Y.: MetaRing – A Full-Duplex Ring with Fairness and Spatial Reuse. IEEE Trans. Commun. 41(1) (January 1993)
3. IEEE Std. 802.5-1989.: EEE Standard for Token Ring
4. Ross, F.E.: Overview of FDDI: The Fiber Distributed Data Interface. IEEE JSAC 7(7) (September 1989)
5. ISO/IECJTC1SC6 N7873: Specification of the ATMR Protocol (v. 2.0) (January 1993)
6. Lemppenau, W.W., Van As, H.R., Schindler, H.R.: Prototyping a 2.4 Gb/s CRMA-II Dual-Ring ATM LAN and MAN. In: Proc. 6th IEEE Wksp., Local and Metro Area Net. (1993)
7. Davik, F., Yilmaz, M., Gjessing, S., Uzun, N.: IEEE 802.17 Resilient Packet Ring Tutorial. IEEE Communications Magazine (2004)
8. Mathieu, C.: Toward Packet OADM. WDM product group, Alcatel-Lucent presentation (December 2006)
9. Chiaroni, D.: Optical Packet Add/Drop Multiplexers: Opportunities and Perspectives. Alcatel-Lucent R&I, Alcatel-Lucent presentation (October 2006)
10. Santitoro, R.: Metro Ethernet Services – A Technical Overview (2003),
    http://www.metroethernetforum.org/
    metro-ethernet-services.pdf
11. Fall, K., Varadhan, K.: The ns Manual. UC Berkeley, LBL, USC/ISI, and Xerox PARC (December 13, 2003)
12. Le Sauze, N., Dotaro, E., Ciavaglia, L., Bouabdallah, N., Dupas, A.: DBORN (Dual Bus Optical Ring Network). An Optical Metropolitan Ethernet Solution
13. Eido, T., Nguyen, D.T., Atmaca, T.: Packet Filling Optimization in Multiservice Slotted Optical Packet Switching MAN Networks. In: Advanced International Conference on Telecommunications, AICT 2008, Athens (June 2008)
14. IP packet length distribution (2002),
    http://www.caida.org/research/traffic-analysis/AIX/
15. Gumaste, A., Chlamtac, I.: Light-Trails: An Optical Solution for IP Transport. Journal of Optical Networking 3(5), 261–281
16. Ramaswami, R., Sasaki, G.: Multiwavelength optical networks with limited wavelength conversion. IEEE/ACM Transactions on Networking (TON) 6, 744–754 (1998)

A Priority-Based Multiservice Dynamic Bandwidth Allocation for Ethernet Passive Optical Networks

Minh Thanh Ngo and Annie Gravey

INSTITUT TELECOM, Telecom Bretagne, Department of Computer Science,
Technopôle Brest Iroise CS 83818, 29238 Brest, France
Université Européenne de Bretagne, France
{mt.ngo,annie.gravey}@telecom-bretagne.eu

Abstract. One of the most cost-effective solutions to meet the explosive increase in bandwidth demand is a Passive Optical Network (PON). This technology is intended to offer a simple, scalable solution which is capable of delivering multiservice access to end-customers. Standards for PONs have been approved, but QoS provisioning and bandwidth distribution within a PON have been left to the implementer. In particular, many proposals have focused on the central issue of upstream traffic management in PONs. Upstream traffic management relies on a Dynamic Bandwidth Allocation (DBA) mechanism that distributes transmission opportunities to the end users, based on their requirements. An ideal DBA should support both Committed Bandwidth (CB) and Best Effort (BE) services, with a good control of the usual QoS characteristics (delay, jitter, and loss) while utilizing the available bandwidth with a good efficiency. The present paper presents a simulation based analysis that shows how a simple priority based DBA meets these requirements, and outperforms several of the better known DBA proposals. *abstract* environment.

Keywords: Ethernet PON (EPON), optical access, quality of service (QoS), dynamic bandwidth allocation (DBA).

1 Introduction

In recent years, an important growth of backbone bandwidth has been observed due to technical developments such as WDM. A similar growth has also been observed in Local Area Networks (LANs) which now routinely support 1Gbps or even 10Gbps interfaces.

On the other hand, the access link is still a bottleneck. This link should be provided by a low cost, simple and scalable access technology, that must support integrated services such as voice, data and video. Access networks relying on optical technologies in terms of Fiber To The Node/Home/Premises/Curb (FTTx) networks, present a very promising improvement over existing access technologies based on copper (xDSL) and cable modem (CATV) for full services broadband access network.

R. Valadas and P. Salvador (Eds.): FITraMEn 2008, LNCS 5464, pp. 185–199, 2009.
© Springer-Verlag Berlin Heidelberg 2009

Broadband Passive Optical Network (PON) is one of the most deployed technologies for delivering FTTx networks. Tree, ring, bus are several topologies suitable for PONs, with the tree topology being the most common architecture.

A PON consists in an Optical Line Termination (OLT) located at the central office (or cable head end) and multiple remote Optical Network Units (ONUs) that deliver broadband voice, data, and video services to subscribers. In PONs, the OLT broadcasts downstream traffic, and ONUs filter frames that are sent to them. In the upstream direction, Time Division Multiplexing (TDM) is used to avoid collisions. Each ONU transmits frames using timeslots assigned by the OLT. The OLT thus acts as the central arbitrator for the transmission of upstream traffic.

Many mechanisms which allow a dynamic allocation of transmission opportunities have been proposed [1], [2]. Those mechanisms assume that the traffic is IP based, and may be multiservice with various QoS requirements. The basic concept of a dynamic allocation of transmission opportunities relies on three features:

1. a signaling mechanism that allows an ONU to describe its needs and the OLT to allocate transmission opportunities;
2. a Dynamic Bandwidth Allocation (DBA) method implemented at the OLT side that computes transmission opportunities offered to each ONU based on negotiated SLAs and dynamic information transmitted by the ONUs;
3. an intra-ONU scheduling that specifies how each ONU uses the allocated grants.

The IEEE 802.3ah [3] standard has specified a Multi-Point Control Protocol (MPCP) for MAC layer of EPON system. This protocol defines a set of messages used to control the data exchange between the OLT and the ONUs: with REPORT messages, the ONUs request transmission opportunities which are granted by the OLT using GATE messages. MPCP is commonly used, and solves point 1. MPCP allows transmitting requests and allocations for several traffic classes.

On the other hand, there is no current agreement regarding points 2 and 3. Although some early proposals such as IPACT [4] are currently considered as benchmarks for other proposals, many issues regarding a multiservice support by EPONs of broadband traffic in the upstream direction are still open.

This paper is focused on the QoS delivered to upstream traffic in EPONs, and on the efficiency of the allocation process. Our preliminary findings led us to design a simple priority based upstream traffic management policy [13]. This policy is further analyzed in the present paper and is compared with other published proposals.

The analysis is performed by simulation, using Network Simulator (ns2). Simulation is indeed mandatory in order to reach a good understanding of the complex interactions between offered traffic, intra-ONU scheduling policies and centralized arbitration by the DBA. Analytical modeling is generally used only to assess global performance indicators such as Mean Queueing Delays [5] and has to rely on drastic simplifications of the exact interactions observed in a PON.

The rest of the article is organized as follow. Section 2 presents an overview of related literature and provides the background knowledge for further discussion. Section 3 presents how a multiservice DBA should ideally behave. Our contribution in terms of dynamic allocation policy is briefly presented in Section 4. Section 5 specifies the simulated network architecture. Section 6 addresses the analysis of the performance delivered by the EPON to upstream traffic and Section 7 concludes the paper.

2   State of the Art in Resource Management in EPON

This section first addresses the design of the polling procedure used in the DBA scheme; as we shall show, it has a significant impact on the performance of the traffic management policy for upstream traffic. We then we provide an overview of some notable research efforts on DBA design for EPONs.

2.1   Discussion of the Polling Scheme

DBA schemes frequently define a so-called "polling cycle". A polling cycle is defined as the time that elapses between two successive transmission windows assigned to the same ONU. The polling cycle can have a fixed or a variable length. Clearly, the polling cycle is an important parameter of the system; it impacts on the QoS that can be supported (both the latency and the jitter increase with longer polling cycles).

The polling procedure used to interrogate the ONUs on their transmission requests and to assign transmission opportunities to them can be designed in many different ways. This is true for all PON architectures, but has a larger impact for EPONs since this technique does not mandate a fixed cycle time; the cycle time can then be variable, and directly depends on the characteristics of the polling procedure.

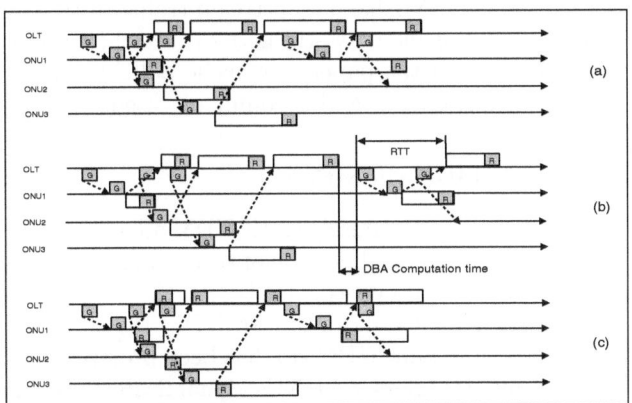

Fig. 1. Polling policies

With poll-and-stop polling, the ONUs are polled and allowed to transmit one after the other, with a complete round-trip time (RTT) between each ONU's transmission periods. Although, due to physical layer requirements, it is indeed mandatory to maintain a guard time (a time during which the upstream channel is idle between two transmission periods by two different ONUs), such a guard time has only to last a few microseconds, whereas a typical RTT is between 100 and 200 microseconds. Poll-and-stop polling obviously wastes upstream bandwidth, especially when the OLT serves a large number of ONUs. This in turn reduces upstream channel utilization and thus increases packet delay.

A more efficient way is allowing the OLT to anticipate and send a GATE messages to an ONU sufficiently before this ONU is allowed to start transmission, so that the ONU can start just after the previous ONU has finished transmitting, with a minimal guard time between transmission periods.

A famous policy [4] is to interleave polling messages: once an ONU has finished transmitting its data, it sends a REPORT, and the OLT replies directly with a GATE message advertising the next transmission period, as shown in Fig1.a. This fulfils the above requirement, except in very pathological cases, when a single ONU is active.

However, Interleaved Polling relies on the OLT allocating transmission opportunities based only on the ONU's request. The OLT is therefore unable to take into account the requests of all ONUs and thus make an more intelligent decision based on the global demands. Fig1.b illustrates an effective way to overcome this drawback that is called Cyclic Polling. In the Cyclic Polling policy, the OLT does not start the next polling cycle before REPORT messages from all ONUs have been received. This allows the OLT to allocate transmission opportunities based on the requests of all ONUs at the end of each polling cycle.

Fig1.b shows that, if the REPORT messages are sent by the ONUs at the end of their transmission periods, the upstream channel may still suffer from under-utilization (i.e. roughly 1 RTT per cycle time). This issue is addressed in [6] by proposing a more efficient cyclic polling method in which ONUs send REPORT messages at the beginning of their transmission period. In some pathological cases (i.e. if the last polled ONU has no upstream traffic to transmit, while the system is congested) this scheme may still underutilize the resources. However, as shown in the remainder of the paper, the cyclic polling policy illustrated in Fig1.c has many major advantages and is commonly used.

2.2   Dynamic Bandwidth Allocation

Considering the shortcomings of a purely static TDM allocation strategy, various mechanisms that allow a dynamic allocation of transmission opportunities in the upstream direction have been proposed.

The first dynamic bandwidth allocation for TDM-PONs, called Interleaved Polling with Adaptive Cycle Time (IPACT), has been proposed for EPON in [4]. IPACT assigns transmission slots to ONUs dynamically. The most common scheduling discipline proposed for IPACT is Limited Service (LS). With LS, the OLT grants the required transmission slots, but no greater than a given predetermined maximum thus limiting the duration of polling cycles. This discipline has been shown to provide good performance in terms of packet delay and throughput for a single traffic class.

Ideal DBA allocation schemes require the OLT not only to fairly and efficiently share the upstream channel, but also to support some type of QoS model. A large variety of QoS-enabled DBA mechanisms have been proposed that differ in the QoS model chosen by their authors.

In [7], the authors propose a DBA scheme (DBA-Assi) in which ONU nodes are partitioned into two groups, "underloaded" and "overloaded", according to their minimum guaranteed transmission window sizes. In that scheme, the bandwidth saved from the underloaded group is proportionally re-allocated to overloaded ONUs. This scheme fairly distributes the excessive bandwidth among highly loaded ONUs. However, the OLT still supports a single traffic class.

In [8], a bandwidth guaranteed polling protocol has been proposed , which allows the upstream bandwidth to be shared based on the service level agreement (SLA) between each ONU and the operator. This protocol is able to provide a bandwidth guarantee for premium ONUs based on SLAs while providing Best-Effort service to other ONUs. However, in this model, an ONU is unable to request different QoS services for different traffic streams.

In [9], the authors propose IPACT-LS-QoS DBA in which every ONU maintains a separate queue for each class of service in its buffer but addresses only "global" requests to the OLT. The OLT issues "global" grants to each ONU, which means that the OLT does not dictate how many bytes exactly from a particular queue the ONU must transmit. Instead, each ONU uses strict priority scheduling to determine the order in which its queues are processed. However, none of the mechanisms proposed in [9] provide bandwidth guarantees for traffic flows specified by SLAs.

Authors in [10] and [11] proposed a "per class subframing" approach to allocate the bandwidth. This approach consists of separating the transmission of high priority and low priority packets into two sub-cycles. By protecting high priority traffic in a separate and fixed sub-cycle, its delay jitter performance is improved. However, "per class subframing" schemes consume upstream bandwidth since each ONU transmits multiple times during each polling cycle, thus increasing the overhead due to guard times t_g.

Authors in [6] also propose a DBA using cyclic polling scheme which supports DiffServ. Expedited Forwarding (EF) traffic is allocated a guaranteed amount of bandwidth regardless of the number of packets waiting in the queue. Assured Forwarding (AF) traffic is allocated bandwidth as per their proportion in the total demand. This DBA precludes however AF traffic to take advantage of inactive EF sources. Moreover [6] does not relate EF and AF classes to typical traffic classes.

As we have briefly reviewed above, the existing DBAs either provide BE support only, or rely on a deterministic evaluation of the requirements for the EF traffic class, assuming that the traffic offered in this class has static requirements. However, this is not true in general: even EF traffic can be bursty, especially in the access network since various demands have yet to be aggregated. It seems therefore necessary to design multiservice DBAs, which can both take into account the global set of demands by the ONUs, and support bursty demands, even for QoS sensitive traffic. Section 4 presents a simple priority based DBA that attempts to fulfill these requirements.

3   Ideal Behavior of a Multiservice DBA

As discussed above, the access network is required to accommodate various kinds of traffic, and in particular, multiservice access is a distinguished feature EPONs are expected to provide. In this section, we first identify a minimal set of traffic classes that can be used to differentiate network services, and then we identify different design goals for a multiservice DBA scheme in EPONs

3.1   Class of Services

In order to efficiently support diverse application requirements, the usual approach in Ethernet networking is to classify packets into a limited number of traffic classes. The standard [12] distinguishes 8 traffic classes but we have chosen to take account of only a limited subset of 3 classes as shown in Tab.I.

Table 1. Traffic types

Classes	Characteristics	Requirements
T_0	Real-time CB	low delay and jitter, limited loss rate, guaranteed bandwidth
T_1	Data CB	limited delay and jitter, limited loss rate, guaranteed bandwidth
T_2	Best Effort	none

Basically, we assume that there are 2 Committed Bandwidth (CB) classes and a single Best Effort (BE) class. A major difference between CB and BE traffic is that CB traffic is characterized by a traffic profile specified in a SLA and expects its QoS requirements to be fulfilled as long as the offered traffic complies with the negotiated traffic profiles. On the other hand, there are no BE traffic profiles and BE traffic cannot request QoS commitments.

3.2   DBA Design Goals

We have identified the following design goals for the design of a DBA scheme in EPONs:

– **Class sensitiveness** : DBA schemes should be QoS-sensitive.

- **Efficiency :** DBA schemes should limit overhead. As discussed in the section 2.1, the choice of polling schemes for DBA impacts directly the upstream channel utilization.
- **Traffic conformance issues:** DBA schemes should take into account traffic conformance issues in order to enforce traffic profiles described in SLAs.

The DBA is a major block of upstream traffic management in EPONs. However, it also has to interact with other features such as dimensioning and/or call acceptance control. However, this is outside the scope of the present paper.

4   A Simple Multiservice DBA

Considering DBA design goals described in the previous section, we have proposed a DBA policy [13], which we called DBA-TCM (Traffic Conformance Mechanism) that has the following characteristics :

- it allows variable length polling cycles while enforcing an upper bound on the maximum polling cycle length T_{max}.
- it is priority based at the inter-ONU level, i.e. the DBA algorithm serves each class successively and attempts to satisfy all the demands from one class before allocating transmission opportunities to a lower class.
- it is priority based at the intra-ONU level, i.e. each ONU applies exactly the allocated per-class grants and frames that arrived after the previous request cannot be served within the current cycle.
- it enforces ONUs to comply with their negotiated SLAs by checking conformance of upstream Committed Bandwidth (CB) traffic: the OLT filters the demands and takes account of a virtual policing scheme when computing the amounts of grants to be sent to the ONUs.

DBA-TCM does not interleave REPORT and GATE messages as IPACT does. Actually, as stated at the end of section 2.1, all REPORT messages received during a given cycle are simultaneously processed by the OLT in order to compute the resources to allocate to each ONU for the next cycle.

A previous paper [13] details the exact algorithm and demonstrates the efficiency of DBA-TCM when dealing with traffic conformance issues (i.e. enforcing SLAs and delivering committed bandwidth to compliant sources). However, the previous paper does not assess the behavior of the DBA relative to QoS issues. This is the objective of this present paper.

DBA and Admission Control

DBA does not protect the system from congestion, it only shares available resources between ONUs. Another paper addresses this issue[14] and shows how an appropriate Connection Admission Control (CAC) policy has to ensure that the global upstream traffic to be carried by the PON can indeed be supported by it. In the following section, we assume that the global offered Committed Bandwidth traffic (T_0 and T_1) volume is small enough to be correctly supported by the PON.

DBA and Fairness

"Fairness" is a target behavior of a system that can control how resources are shared between users. Obviously fairness only applies when the total amount of resources is not sufficient to accomodate all demands. In [13], we state that the PON operators can reserve bandwidth for Best-Effort traffic. In that case, even if ONUs submit more Committed Bandwidth traffic than can be supported in a polling cycle, some BE traffic is transmitted. However, in the present paper, we assume that no ressources are dedicated to BE traffic, the later is only served when all CB traffic is accomodated.

5 Simulating EPON Traffic Management

The simulated network architecture and the various traffic types that are considered in the following analysis are now specified. The simulation platform used for this work has been developed in our team and is described in [15].

Network Architecture

We consider an EPON system with 16 ONUs, and a total uplink bandwidth of 1.0 Gbps. The maximum cycle time T_{max} is set to 1.5 ms and the guard time t_g between 2 consecutive ONU transmissions is 4 μs.

The exchange of GATE and REPORT messages is not simulated. On the other hand, the simulation tracks the level of each ONU buffer, and the OLT bases its computations on the values taken by buffer levels at the time when the REPORT message would have been transmitted by the ONU.

Traffic Modeling

In our experiments, class T_0 is simulated as a real-time VBR stream which is represented as an ON-OFF model. The duration lengths D_{on}, D_{off} of an ON and OFF period are generated according to an exponential distribution with a same mean of 10 ms. Class T_1 is also represented as an ON-OFF exponential model, ON and OFF average duration periods are both set to 200 ms. In ON-OFF model, the average rate R_a can be obtained by

$$R_a = \frac{D_{on}}{(D_{on} + D_{off})} R_p \tag{1}$$

where R_p is the rate at which a source during ON state transmits. Class T_2 represents the best-effort traffic. Each ONU may have different traffic profiles specifying T_0 and T_1 traffic flows. Depending on the scenario, T_2 traffic is modeled either as a Pareto source, or as a greedy (i.e. always active) FTP source.

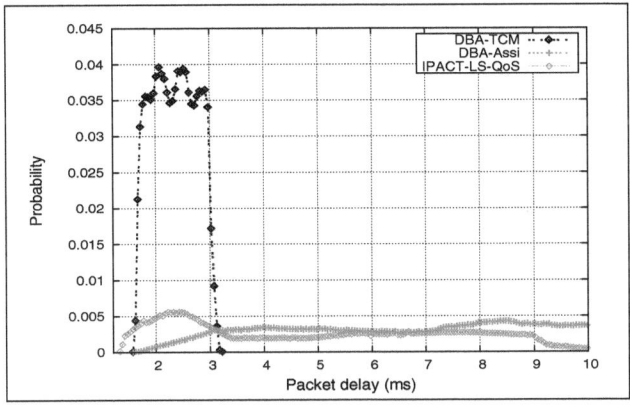

Fig. 2. Impact of a class sensitive DBA on QoS for class T_0

6   QoS Delivered to Upstream Traffic

In this section, we first show that both the centralized computation of transmission opportunities and the intra-ONU scheduling policy should be QoS sensitive. We then analyze the efficiency of DBA-TCM, showing that BE traffic can fill the capacity of the upstream traffic while the committed QoS is delivered to CB traffic. Lastly, the impact of the polling procedure on delivered QoS is addressed.

6.1   Class Sensitiveness Issues

The DBA-TCM described in section 4 is class sensitive: it serves the demands in priority order. The following scenario justifies this design choice.

In this scenario, 4 ONUs send only T_0 traffic (R_a=30 Mbps), 12 others ONUs send T_2 traffic (greedy FTP sources). Fig.2 represents the T_0 packet delay distribution for 3 DBAs: IPACT-LS-QoS[9], Assi[7] and DBA-TCM described in section 4.

Fig.2 shows that the delay for DBA-TCM presents centralization with all data points condensed before 3.0 ms whereas the two others present a heavier tail. In other words, only DBA-TCM provides a very short delay for T_0 traffic, while both IPACT-LS QoS and Assi distributions present very long tails.

This is easily explained: IPACT-LS QoS applies an Inter-Leaved Polling scheme in which the OLT does not have an overview of all demands from the ONUs when computing transmission opportunities. Therefore, high-priority traffic is handled by the OLT in the same manner than BE traffic, which results in a serious degradation of the QoS delivering to T_0 traffic as illustrated in Fig.2. The DBA scheme proposed by Assi experiences also a T_0 traffic QoS degradation. Indeed, in this scheme, the OLT still only focuses on how to satisfy bandwidth requests from different ONUs by fairly distributing the excessive bandwidth among highly loaded ONUs and does not provide a better service to high priority traffic.

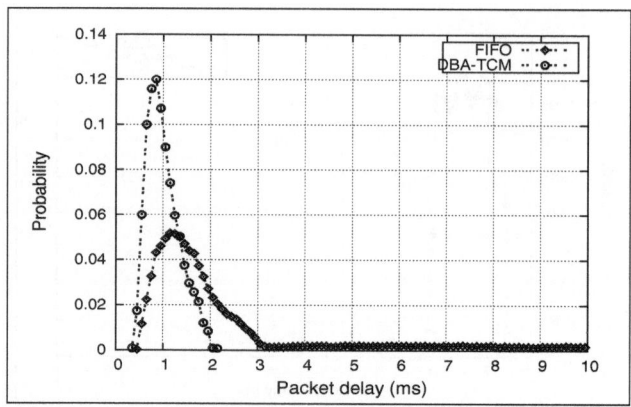

Fig. 3. Impact of FIFO intra-ONU scheduling on QoS for class T_0

Taking QoS into account at the intra-ONU scheduling level is also necessary as shown in the following scenario. In this scenario, we assume that all ONUs generate T_0 traffic (R_a=30 Mbps) plus some T_2 traffic (a Pareto source with a 5 Mbps mean rate). In the first case, the intra-ONU scheduling policy is pure FIFO, serving the packets in the order of their arrivals, whereas in the second case, the intra-ONU scheduling policy is not FIFO, but serves each traffic class as specified in the GATE messages. It may then happen that T_0 overtakes T_2 traffic.

Fig.3 represents the T_0 packet delay distribution in the two cases. It clearly shows the impact of the FIFO mechanism on the QoS delivered to CB traffic: the peak is much flatter and the tail heavier. Obviously, the presence of T_2 traffic, even in small proportion, degrades the QoS delivered to the T_0 traffic.

These preliminary studies justify the design choice of treat BE and CB traffic separately both for inter-ONU scheduling and intra-ONU scheduling.

6.2   Analysis of Upstream Channel Utilization

Good upstream channel utilization is a design objective of EPON systems. This utilization is very dependent on the DBA.

We first compare the channel utilization delivered by several DBA schemes: a static TDMA scheme, the IPACT-LS with a standard maximum window size of 15000 bytes)[9] and the DBA TCM.

The simulation scenario is as follows: only one ONU out of 16 is active, and sends BE T_2 traffic. The T_2 rate of this ONU varies from 50 Mbps to 1 Gbps during the simulation. Although this is indeed an extreme case, it is representative of what happens when a single heavy user is on the net, while the other ONUs are inactive. Fig.4 shows the upstream channel utilization for the 3 DBAs under study versus the offered T_2 traffic load.

The simulation shows that the TDM static policy achieves only 6% maximum utilization compared with 60% for IPACT-LS and with 95% for DBA TCM. The

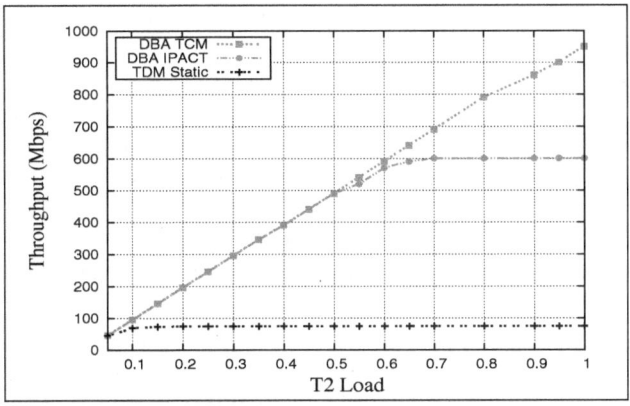

Fig. 4. Impact of polling policies on channel utilization

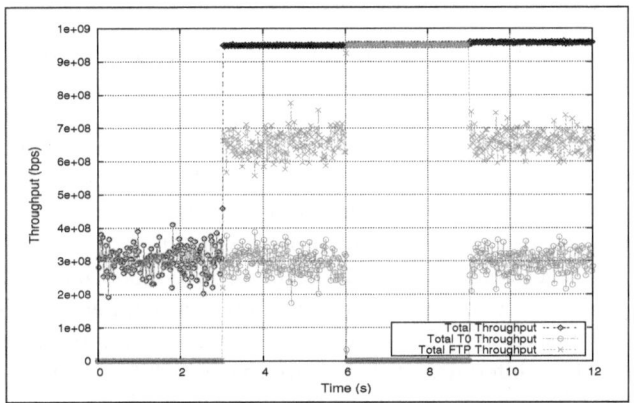

Fig. 5. Upstream channel utilization using DBA TCM

TDM static scheme achieves the worst channel utilization because each ONU is given a fixed transmission window regardless of the ONU's activity.

Compared to the static TDM case, IPACT-LS does not grant transmission opportunities to idle ONUs. In doing so, IPACT-LS achieves a much better channel utilization. However, IPACT-LS blindly limits the number of grants allocated to one ONU, irrespective of the activity of the other ONUs. This is why the maximum utilization is 60% (this limit depends on the maximum window size). On the other hand, since DBA-TCM knows about all requests, it can allocate the full bandwidth to the single active ONU and thus achieve a maximum utilization.

In order to analyze a less extreme case, we consider a scenario in which more ONUs are active and vary their respective activity levels. In this particular scenario, T_0 sources (R_a=18.75 Mbps) in each of 16 ONUs are active only from

0.0 s to 6.0 s and from 9.0 s to 12.0 s. Each ONU also starts a FTP session at 3.0 s. Upstream traffic is controlled by DBA-TCM.

Fig.5 represents the achieved utilization versus time. In the first 3.0 s, only T_0 traffic is carried and the system is underloaded since the offered traffic corresponds roughly to 30% the capacity of the system. When the FTP sources are activated, the upstream channel is almost full, since FTP traffic grabs all available bandwidth (the channel is not completely full, due to the guard time overhead). Between 3.0 s and 6.0 s, the capacity used by T_0 traffic is unchanged while FTP grabs more resources between 6.0 s and 9.0 s when all T_0 sources are idle.

The main point is that, as long as T_0 traffic is sent, it receives the expected throughput, whether or not FTP sources are activated. These results clearly demonstrate the efficiency of DBA-TCM in terms of ensuring high upstream channel utilization while delivering the QoS commitments to CB traffic.

6.3   Impact of Polling on Delivered QoS

All DBA policies implement mechanisms to limit the latency of the system at high load. Indeed, it may happen that the total offered traffic, over a given period of time, exceeds the capacity offered by the PON to upstream traffic. In order to avoid a global degradation of the performance, the DBA has to ensure isolation between ONUs, i.e. offer a good latency to ONUs that do not send excess traffic.

Since upstream traffic management is based on polling, the polling cycle, which in EPON is variable, should be upper-bounded.

In IPACT, as described in [9], no ONU can receive more than a given amount of resources per cycle. This automatically enforces a maximum cycle time.

As explained in section 4, DBA-TCM also enforces an upper bound T_{max} on the cycle time. The difference with the approach in [9] is that the limit T_{max} is enforced globally and not on a per ONU basis. This is made possible by the global computation and allocation process performed by the OLT for each cycle.

Let us now assess the impact of T_{max} on the latency offered to upstream traffic.

Two scenarios are compared. In both scenarii, ONUs send identical and balanced traffic. In the first scenario, only T_0 traffic is sent, and the total offered load varies between 100.0 Mbps and 800.0 Mbps. In the second scenario, we add an active FTP source in each ONU, which is transmitted as T_2 traffic. As in the previous cases, T_{max}=1.5 ms.

Fig.6 shows an upper quantile of the T_0 packet delay versus total T_0 load in the system. This upper quantile represents, as usual, the maximum delay performance delivered to the T_0 traffic.

We can see in Fig.6 that, as expected, the maximum T_0 packet delay increases with the offered T_0 load, and the latency is worse in the second scenario than in the first. On the other hand, an unexpected behavior is also observed : at low T_0 traffic load, the delay performance offered to T_0 traffic is significantly worse in the case when there is T_2 traffic, whereas when the T_0 traffic load is large enough, the delay performance offered to T_0 traffic appears not to be impacted

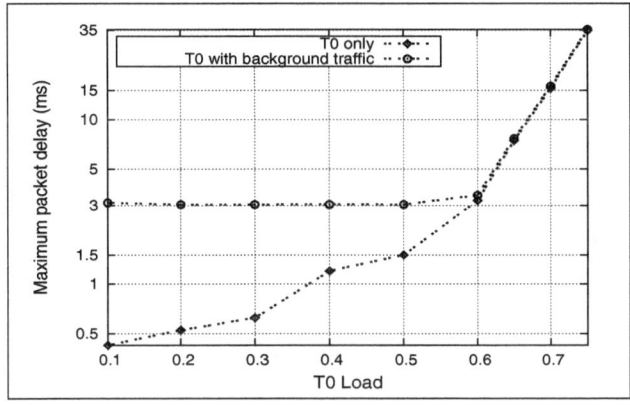

Fig. 6. Impact of total T_0 load on QoS for class T_0

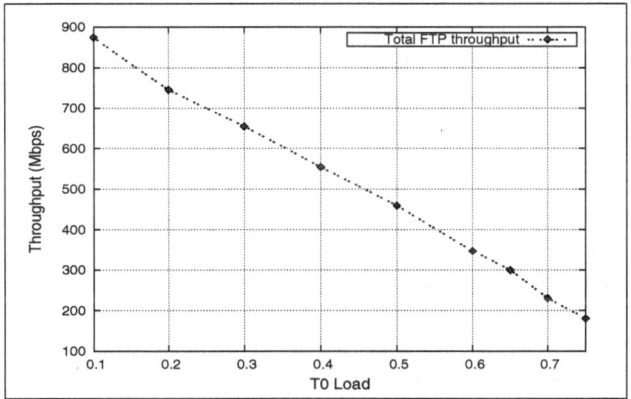

Fig. 7. Total FTP throughput in function of T_0 load

by the presence of T_2 traffic. The limit between the two zones corresponds to a delay performance of 3 ms, i.e. twice the maximum cycle time.

This (apparently) strange phenomenon is explained as follows: although T_2 traffic has less priority than T_0 traffic, it can be served as long as the cycle time due to T_0 traffic only is smaller than the maximum cycle time set to 1.5 ms. When T_2 is served, it increases the cycle time, and thus degrades the QoS offered to T_0 traffic. On the other hand, when the amount of T_0 traffic is large enough, some cycles reach the cycle time limit of 1.5 ms , with only T_0 traffic, and almost no T_2 traffic is served in those cycles. The maximum delay for T_0 traffic, when the cycle times are limited to 1.5 ms due to T_2 traffic, is then equal roughly to 3 ms since the worst case for a T_0 packet is to arrive at an ONU just after a REPORT was sent. In that case, it has to wait 2 cycles, i.e. 3 ms. On the other hand, this is a worst case and in some cycles, T_2 traffic is indeed served as is shown in Fig. 7 that shows the achieved T_2 throughput versus T_0 offered load.

Selecting T_{max} thus implies a trade-off between efficiency and latency. Indeed, a small T_{max} improves the latency as seen above. On the other hand, a large T_{max} increases the efficiency since it decreases the total amount of overhead (guard times between transmissions by different ONUs). Moreover, the larger the T_{max}, le smaller is the CPU used by the OLT to perform resource allocation.

7   Conclusion

This paper has presented a simulation based performance analysis of EPON upstream traffic management policies. Typical issues which have been addressed: class sensitive support, priority handling, channel utilization and QoS differentiation.

All these issues have been analyzed for DBA-TCM proposed in [13].

For the moment, DBA-TCM supports a single Best Effort traffic class, and two Committed Bandwidth traffic classes, including one intended for time sensitive applications.

The results of the present paper show that the DBA-TCM favorably compares with other published proposals by ensuring a very high utilization while supporting QoS differentiation in a simple and efficient manner. DBA-TCM can thus be used to easily offer a combined support of real time and interactive data services, while authorizing BE traffic to use remaining bandwidth.

To the authors knowledge, DBA-TCM is the first proposed DBA that simultaneously offers a multiservice support and an efficient use of available resources by Best Effort traffic.

We have shown that the efficiency of DBA-TCM lies in the fact that resource allocation is performed by the OLT, which centralizes all ONUs' requests on a per cycle basis, and can then optimally distributed the available resources. This principle allows easily extending DBA-TCM, for example by increasing the number of traffic classes. This number could easily be increased by e.g. increasing the number of priority levels. Another possible extension would implement some type of fair scheduling for different classes. This would allow e.g. ensuring a minimal support for Best Effort Traffic even when the amount of submitted Committed Bandwidth traffic is large. The centralized scheduling algorithm in DBA-TCM can thus be used to implement functions such as per class/per ONU conformance control and/or fair scheduling.

Acknowledgments. The work described in this paper was carried out with the support of the French "Agence Nationale de la Recherche" in the framework of the ECOFRAME project (grant number 2006 TCOM-002-06).

References

1. Zheng, J., Mouftah, H.T.: Media Access Control for Ethernet Passive Optical Network: An Overview. IEEE Communications Magazine 43(2), 145–150 (2005)

2. McGarry, M., Maier, M.P., Reisslein, M.: Ethernet PONs: a survey of dynamic bandwidth allocation (DBA) algorithms. IEEE Communications Magazine 42(8), 8–15 (2004)

3. Ethernet in the First Mile Task Force IEEE Std 802.3ah (2004), http://www.ieee082.org/3/efm/

4. Kramer, G., Mukherjee, B., Pesavento, P.: Interleaved Polling with Adaptive Cycle Time (IPACT): A Dynamic Bandwidth Distribution Scheme in an Optical Access Network. Photonic Network Communications 4(1), 89–107 (2002)

5. Ngo, M.T., Bhadauria, D., Gravey, A.: A mean value analysis approach for evaluating the performance of EPON with Gated IPACT. In: Proc. IFIP/IEEE ONDM 2008, Barcelona, Spain (2008)

6. Choi, S.-i.: Cyclic Poilling Based Dynamic Bandwidth Allocation for Differentiated Class of Service in Ethernet PONs. Photonic Network Communications 7(1), 87–96 (2004)

7. Assi, C.M., Ye, Y., Dixit, S.: Dynamic bandwidth allocation for quality-of-service over Ethernet PONs. IEEE J. Sel. Areas Commun. 21(9), 1467–1477

8. Ma, M., Zhu, Y.: A bandwidth guaranteed polling MAC protocol for Ethernet passive optical networks. In: Proc. IEEE INFOCOM, San Francisco, CA, USA, vol. 1, pp. 22–31.

9. Kramer, G., Mukherjee, B., Pesavento, P.: On supporting differentiated classes of service in EPON-based access network. J. Opt. Networks, 280–298 (2002)

10. An, F.T., Hsueh, Y.: A new dynamic bandwidth allocation protocol with quality of service in ethernet-based passive optical networks. In: Proceeding of WOC 2003 (2003)

11. Shami, A., Bai, X., Assi, C.M., Ghani, N.: Jitter Performance in Ethernet Passive Optical Networks. J. of Lightwave Technology 23(4), 1745–1753 (2005)

12. ANSI/IEEE Standard 802.1D, http://www.standards.ieee.org/

13. Ngo, M.T., Gravey, A.: Enforcing Bandwidth Allocation and traffic conformance in Passive Optical Networks. In: Proc. ICST/IEEE AccessNets 2007, Ottawa, Canada (2007)

14. Ngo, M.T., Gravey, A.: A Versatile Control Plane for EPON-based FTTX Access Networks (submitted for publication)

15. Ngo, M.T., Gakhar, K., Gravey, A.: Upstream Traffic Management in EPONs: A Simulation Based Analysis. In: Proc. ICST Simutools 2008, Marseille, France (2008)

High-Performance H.264/SVC Video Communications in 802.11e Ad Hoc Networks

Attilio Fiandrotti, Dario Gallucci, Enrico Masala, and Juan Carlos De Martin

Politecnico di Torino, Dipartimento di Automatica e Informatica
Corso Duca degli Abruzzi 24, 10129 Torino, Italy
{attilio.fiandrotti,dario.gallucci,enrico.masala}@polito.it,
juancarlos.demartin@polito.it

Abstract. This work focuses on improving the performance of video communications based on the recently developed H.264 Scalable Video Coding (SVC) standard over 802.11e wireless networks. The H.264/SVC standard is particularly suitable for wireless communications because of its compression efficiency and the ability to adapt to different network scenarios. The adoption of the H.264/SVC codec in 802.11 networks poses however some issues. In particular, since pictures are encoded into several small video units, the overhead imposed by the 802.11 contention-based channel access mechanism might be large. Thus, the strategy employed for the packetization of the video data plays a key role in determining the performance of the network. This work proposes two network adaptation strategies for H.264/SVC video to efficiently use the QoS-enabled 802.11e extension of the 802.11 standard by designing a scheme for joint optimization of video data aggregation and unequal error protection. Simulations of video transmissions in a realistic home networking scenario characterized by direct communications between devices in ad hoc mode show that the proposed strategies reduce the packet loss rate and significantly improve the quality of the communication with PSNR gains up to about 2 dB. Moreover, the performance of the proposed low-complexity strategy is close to that of the optimal, high-complexity, strategy.

1 Introduction

Scalable video coding is increasingly used in multimedia communications since the recently developed H.264 scalable video coding (SVC) standard [5] achieves nearly the same compression efficiency of the state-of-the-art non-scalable H.264 advanced video coding (AVC) standard. A scalable video stream consists of a low-bandwidth substream (base layer) which represents a low-quality version of the video (e.g., low temporal o spatial resolution) and one or more substreams (enhancement layers) which, together with the base layer, allow the reconstruction of the full quality video. Scalability allows the transmission of the same

R. Valadas and P. Salvador (Eds.): FITraMEn 2008, LNCS 5464, pp. 200–210, 2009.
© Springer-Verlag Berlin Heidelberg 2009

encoded bitstream over different types of networks to client devices with different processing capabilities. Enhancement layers, in fact, can be dropped in case of network congestion or if the client is unable to decode them due to, e.g., processing power constraints. Therefore, scalability is particularly valuable in communications over wireless networks, where neither constant link quality nor minimum bandwidth can be easily guaranteed due the intrinsic unreliability of the medium.

The increasing popularity of multimedia-capable devices equipped with an 802.11 wireless network interface is also fostering the development of more and more efficient wireless video communication schemes. Moreover, QoS-enabled extensions such as the 802.11e [1] can be exploited in conjunction with the unequal perceptual importance of multimedia data to design error protection schemes specifically for video. For instance, Ksentini [6] exploited the different access categories provided by the 802.11e protocol to grant better protection to the most important parts of AVC streams. More recently, the work in [8] showed that, given the same error protection scheme, the SVC codec provides better video quality than AVC since it better adapts to different network conditions.

SVC communications over 802.11-based networks raise however some new peculiar issues with respect to traditional AVC communications due to the high number of channel accesses they require. Such issues are especially evident in ad hoc scenarios, where the absence of a centralized infrastructure does not allow to coordinate and optimize channel usage. The decomposition of each picture in base and enhancement information results in fact in a large number of packets offered to the network in the common case that each information unit is transmitted as a new packet, as recommended by the RFC standard [3]. Transmitting a high number of small packets on a 802.11 network performs worse than transmitting a small number of large packets due to the overhead associated with channel access contentions [11]. This issue also affects VoIP applications, limiting the number and the quality of concurrent communications that the network can sustain.

While much attention has been devoted to the case of VoIP communications (see, e.g., [10]),video communication has received less attention. In this paper we propose and test two strategies which mitigate the channel contention overhead by aggregating multiple video information into a single packet while taking into account the varying perceptual importance of video data. The performance of the proposed strategies is evaluated by simulation in an 802.11e network scenario using both network and application level metrics. To the best of our knowledge, no previous work has explored the impact of aggregating SVC video data in terms of application level metrics such as the perceived video quality.

The paper is organized as follows. Sec. 2 provides the technical background, followed, in Sec. 3, by the proposal of possible strategies to improve the communication performance. After simulation setup (Sec. 4), results are reported in Sec. 5. Conclusions are drawn in Sec. 6.

2 Background

2.1 The H.264 Scalable Video Coding Standard

The H.264/AVC standard splits the functionalities of the encoder between the Video Coding Layer and the Network Abstraction Layer (NAL) [7] components. The former layer encompasses all the encoder core functionalities such as macroblock encoding. The latter layer facilitates the transport of video over packet networks by encapsulating each piece of encoded data into an independently decodable transport unit known as NAL Unit (NALU). Each NALU is prefixed by a header which specifies the type of data that the NALU contains, e.g., picture slices, parameter sets, etc. As in other video compression standards, the dependencies among the pictures are constrained within the so called *Group of Pictures* (GOP). Figure 1 depicts a typical AVC GOP: each box represents a NALU, the letter inside the box the picture type (Intra, Predictively or Bipredictively coded), the subscript number display order of the picture and the arrows show the decoding dependencies.

The H.264/SVC amendment extends H.264/AVC with temporal, spatial and fidelity scalability options, encoding a video as an independently decodable base layer and one or more enhancement layers. The header of an SVC NALU is extended with extra fields, such as the Temporal Index (TID) field, which provide information about the type of the enhancement information and their level in the decoding hierarchy. Figure 2 depicts the GOP structure of a typical H.264/SVC

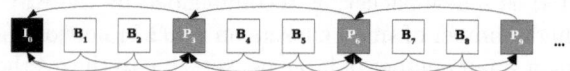

Fig. 1. A typical H.264/AVC GOP structure

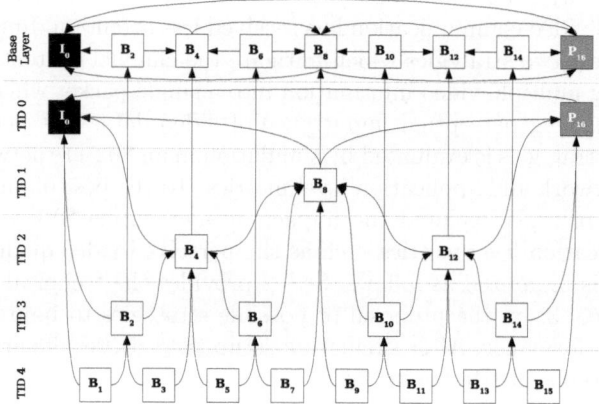

Fig. 2. GOP structure of the H.264/SVC encoding scheme used in this work

encoding scheme which encompasses the AVC-compatible base layer, one spatial and one temporal SVC enhancement layers. Each GOP is 16 frames long and every 32 frames a picture is Intra-coded, therefore the video encoding scheme is $I_0B...BP_{16}B...BI_{32}$.

From Fig. 1 and 2 it is clear that the SVC codec encodes each GOP in a higher number of NALUs and employs a more complex prediction scheme than AVC. The base layer is used to predict some enhancement NALUs while the other NALUs with a higher TID value depends on those with lower TID value as indicated by the arrows. In case a NALU is lost, different error propagation paths are possible. The loss of any base layer NALU generates distortion which affects both the base and the enhancement layers. The loss of an enhancement NALU, instead, generates distortion which affects the enhancement layers with a higher TID index only. Therefore, an efficient protection strategy might protect each NALU according to its potential to propagate distortion to other pictures.

2.2   The 802.11 and 802.11e Communication Standards

The 802.11 standard supports unstructured communications by means of the Distributed Coordination Function (DCF), which is based on a Carrier Sense Multiple Access with Collision Avoidance (CSMA/CA) channel access mechanism. At each host, packets expecting transmission are enqueued whilst the channel is busy. When the medium becomes idle, the host waits for a time interval called DCF Inter Frame Space (DIFS) and, if the medium is still idle, the hosts defers the transmission for the so called backoff interval, i.e., an additional randomly chosen number of transmission slots. Then, if no transmission on the channel is detected, the host transmits the packet and waits for an acknowledgment from the destination. Should noise on the channel or a packet collision make the frame undecodable, the sender would receive no acknowledgment. Thus, the transmitter starts again the transmission of the packet increasing the contention window (CW), which leads, on average, to longer transmission delays.

The recently standardized IEEE 802.11e amendment [1] introduces the Enhanced Distributed Channel Access (EDCA) mechanism which extends the DCF mechanism with QoS support. For each host station, four distinct transmission queues known as Access Categories (ACs) are introduced in place of the unique queue offered by the 802.11 standard. The CW and DIFS values differ from queue to queue, resulting in high priority and low priority queues characterized by unequal chances of getting access to the channel. As a result, packets in high priority queues are elected for transmission before others, resulting in an effective intra and inter hosts traffic prioritization mechanism.

The efficiency of the contention-based DCF/EDCA mechanisms decrease when the number of channel accesses increases due to, e.g., a higher number of packets to send or an increase of the number of hosts participating in the network, since that packets have to wait in queue more time before being transmitted. Moreover, a high number of channel accesses by multiple hosts leads to a higher collision probability and higher packet loss rates.

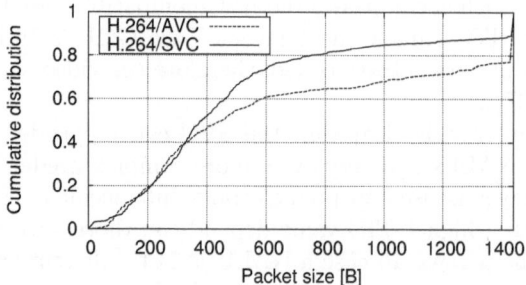

Fig. 3. Cumulative packet size distribution for the Foreman sequence encoded in H.264/AVC and H.264/SVC format. The transmission of an SVC video implies the transmission of a higher number of smaller packets with respect to the AVC video.

Previous work showed that the DCF/EDCA performance can be significantly improved by reducing the number of accesses to the channel, e.g., by transmitting many data units as a single large packet. The 802.11 MAC mechanism is indeed provided with a mechanism which can aggregate small packets in the MAC queue. This mechanism is however designed for generic data and does not take into account application layer constraints such as, for example, the maximum allowed delay. Moreover, such mechanism cannot discriminate among the different types of multimedia streams nor it does consider the different perceptual importance of the various parts of a multimedia stream.

3   Problem Analysis and Solution

Generally, a video sequence encoded in SVC format requires the transmission of a higher number of RTP packets compared to a non-scalable AVC sequence. According to the SVC scheme depicted in Figure 2, every odd-numbered picture (namely the first, the third, etc.) is encoded as one base and one enhancement NALU, that is, on average, each two pictures three NALUs are generated. On the contrary, in the case of the AVC codec each picture can be encoded a single NALU (Figure 1). When the encoded video is transmitted over an IP network, each NALU is encapsulated into a single RTP packet (unless fragmentation is required) according to the RFC draft [3]. The higher number of NALUs of an SVC video thus implies the transmission of a higher number of packets with respect to a non scalable AVC video.

In this work a set of well known test video sequences, each 9~10s long, is encoded using the AVC JM [12] and the SVC JSVM [13] reference encoders provided by the ISO/MPEG Joint Video Team. Table 1 reports the bitrate, the encoding PSNR, the number and the average size of the encoded VCL NALUs as well as the number and the average size of the resulting RTP packets. A sample of the RTP packet size distribution for the Foreman sequence is also shown in Figure 3 for both the AVC and SVC formats. The figures confirm that

Table 1. Characteristics of the test video sequences encoded in H.264/AVC and H.264/SVC format

	H.264/AVC						H.264/SVC					
Sequence	PSNR [dB]	Bitrate [kb/s]	# of NALUs	NALU size	# RTP pkts.	RTP pkt. size	PSNR [dB]	Bitrate [kb/s]	# of NALUs	NALU size	# RTP pkts.	RTP pkt. size
Coastguard	30.57	326	298	1361	499	825	30.28	337	450	934	600	703
Football	31.34	615	259	2562	604	1111	31.07	571	390	1583	724	867
Foreman	33.29	205	298	854	386	672	33.13	212	450	586	522	521
Soccer	32.47	302	298	1259	425	895	32.98	331	390	915	476	766
Tempete	30.63	365	259	1523	432	926	30.46	377	390	1045	536	775

an SVC video communication implies the transmission of a higher number of smaller RTP packets with respect to its AVC counterpart, with the drawbacks highlighted in Section 2.2.

In this work we propose and test two different algorithms to reduce the number of channel accesses required by an SVC communication. Both algorithms aggregate multiple NALUs into a single RTP packet on a GOP basis, according to the RFC draft [3], therefore creating a standard-compliant packet flow. The I-type and P-type NALUs usually exceed in size the network MTU, therefore they cannot be aggregated together. On the contrary, the B-type enhancement NALUs with high TID index are good candidates for aggregation since they are copious and their size is much lower than the MTU.

The first proposed algorithm is referred to as *TID-based* in the rest of the paper and aggregates NALUs with identical TID in decoding order unless the network MTU (or a desired RTP packet size threshold) is exceeded. The number of NALUs which can be encapsulated into a single packet is highly variable and depends on the MTU, on the threshold imposed on the RTP packets size and on the video sequence. With reference to Figure 2, the proposed strategy would aggregate NALUs B_4 and B_{12} into a single RTP packet, B_2, B_6, B_{10} and B_{14} into a single packet and $B_1 \ldots B_{15}$ together. It is however unlikely that all the enhancement NALUs with the same TID can be aggregated into the same RTP packet given the MTU and the constraint on the maximum packet size. For example, NALUs B_2 and B_6 will be aggregated in a single packet as well as NALUs B_{10} and B_{14}, depending on their size. After aggregation, existing prioritization schemes for SVC video based on the TID value [8] can be easily applied.

The second algorithm is referred to as *Optimal* and aggregates NALUs regardless of the TID. The algorithm is optimal in the sense that, by means of an exhaustive search, it finds the combination of NALUs that minimizes the number of packets offered to the network. Therefore its computational complexity is high and in this work is used mainly as a reference for the *TID-based* algorithm. Since the *Optimal* strategy may aggregate NALUs with different TID into a single packet, the issue of determining the perceptual importance of such packets arises. In this work the importance $T(p)$ of a packet p is determined as shown in Eq. (1):

$$T(p) = \frac{\sum_i T(i) \cdot S(i)}{\sum_i S(i)}. \tag{1}$$

$T(p)$ is the weighted average of the TID values $T(i)$ of each packet i, the weights being the packet sizes $S(i)$, excluding the RTP header overhead. The TID of base layer NALUs is assumed to be equal to -1, thus packets transporting perceptually important NALUs will have their $T(p)$ value close to zero if not negative. Clearly, lower TID values represents higher importance. For example, a 1200-byte packet which includes an 800-byte base layer NALU and a 400-byte enhancement NALU with TID equal to 3, has a $T(p)$ value equal to 0.33. The $T(p)$ values are computed for each packet, and then the same traffic prioritization approach used for the TID-based strategy can be applied.

4   Simulation Setup

The two proposed NALU aggregation strategies are evaluated using the *ns* [4] network simulator. Figure 4 illustrates the network setup, which represents a typical domestic environment where different types of hosts, each located in a different room of a building, compete for the access to the channel. Host A is the gateway which provides internet access to all the other hosts in the building. Host G is a TV box which receives a multimedia stream composed of an SVC video and an AAC audio substreams from a content provider located in

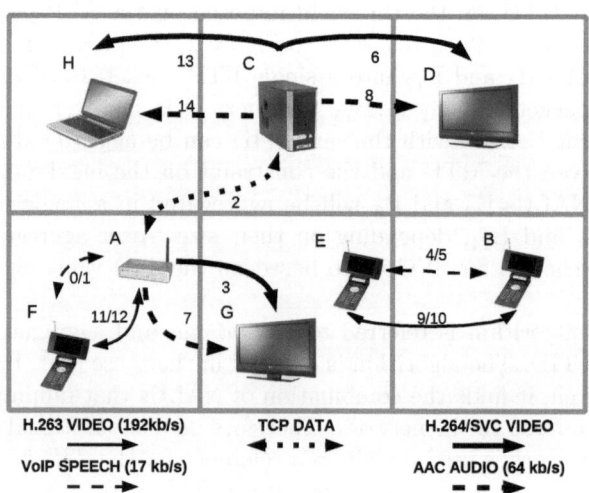

Fig. 4. The simulated network topology

the internet. Host F is a videoconferencing device which communicates with a remote host located in the internet by sending and receiving VoIP traffic and low bitrate H.263 video. Hosts B and E are videophones which communicate among themselves, each generating traffic whose characteristics, in type and bandwidth, are similar to the ones of node F. Host C is a PC which exchanges data with a host on the internet via a TCP connection, and it additionally acts as a domestic media server which streams AAC audio and SVC video to hosts D and H.

In such a scenario data flows with different bandwidth and delay requirements coexist. For example, a VoIP call requires limited bandwidth, although it loads the network with a high number of small packet which have tight maximum delay and jitter requirements. Videoconferencing traffic demands more bandwidth than VoIP and, similarly, requires timely packet delivery. Maximum delay requirements for streaming of pre-recorded contents are less stringent, albeit a minimum bandwidth is required to ensure a smooth playback. Therefore, the traffic which loads the network is categorized in four classes with different QoS requirements. The highest importance Class A encompasses the four VoIP streams. Class B includes the four H.263 video streams and the 50% perceptually most important traffic, i.e., approximately all the base layer of the three SVC flows. Class C encompasses the three AAC streams and the 25% of the SVC enhancement traffic, i.e., the part with low TID values. Finally, the background TCP traffic and the remaining SVC enhancement traffic are assigned to the lowest importance Class D. The four traffic classes defined in the 802.11e standard are exploited by mapping Class A, B, C and D to AC3, AC2, AC1 and AC0, respectively.

In our simulations the 802.11e extension for *ns* developed by the Berlin Technische Universität [2] is used and the bandwidth of the channel is set to 11 Mb/s. The 802.11e MAC is modified adding a timeout mechanism to the transmission queues which drops a packet after 0.5 seconds of stay in the queue, as recommended by the 802.11 standard. Each SVC bitstream is fed to a RTP packetizer which implements the two NALU aggregation algorithms described in Section 3. Two sets of simulations evaluate the performance under different maximum RTP packet size constraints (750 and 1350 bytes). The same traffic trace is used to generate data flows #3, #6 and #13, namely, the three SVC flows. The video sequences are decoded using a frame copy error concealment technique, then they are visually inspected and PSNR is computed. Both error free and noisy channel conditions ($2.5 \cdot 10^{-5}$ byte error rate at the physical level) are simulated.

5   Results

Table 2 shows the performance of the two NALU aggregation strategies (*TID-based* and *Optimal*) presented in Section 3 both in error-free and noisy channel conditions (respectively, upper and lower half of the table). The table reports the number of SVC packets offered to the network, the byte loss rate (BLR) for the video flow at the application level, the packet triptime and the PSNR of the received video (flow #3). The number of packets offered to the network is

Table 2. Performance of the two NALU aggregation strategies

	Sequence	No aggregation				Max pkt. size [B]	TID-based strategy				Optimal strategy			
		# RTP pkts.	BLR [%]	Trip time [s]	PSNR [dB]		# RTP pkts.	BLR [%]	Trip time [s]	PSNR [dB]	# RTP pkts.	BLR [%]	Trip time [s]	PSNR [dB]
Error-free channel	Coastguard	617	9.91	0.65	27.94	750	556	7.45	0.62	28.14	448	2.60	0.59	29.29
						1350	491	2.88	0.49	28.62	359	1.08	0.47	29.93
	Football	724	28.09	0.86	21.70	750	720	28.67	0.86	21.56	647	27.02	0.84	22.82
						1350	704	28.11	0.85	21.82	601	27.39	0.87	23.36
	Foreman	522	5.45	0.55	32.15	750	447	1.51	0.39	32.76	338	0.08	0.31	33.10
						1350	382	0.03	0.22	33.11	237	0.00	0.17	33.13
	Soccer	477	8.41	0.60	30.64	750	472	7.62	0.58	30.82	441	5.72	0.59	31.44
						1350	416	2.88	0.47	32.08	322	1.03	0.43	32.65
	Tempete	536	12.22	0.64	28.36	750	523	10.45	0.62	28.98	469	9.20	0.65	28.80
						1350	453	6.17	0.57	28.10	342	1.38	0.47	29.99
Noisy channel	Coastguard	617	17.69	0.82	26.22	750	556	15.51	0.82	26.06	448	10.61	0.80	27.92
						1350	491	11.83	0.67	26.46	359	6.81	0.75	28.61
	Football	724	36.70	1.07	18.84	750	720	37.19	1.09	18.70	647	34.99	1.07	20.36
						1350	704	37.21	1.06	19.33	601	36.82	1.11	21.17
	Foreman	522	9.99	0.70	31.22	750	447	4.65	0.59	31.95	338	2.02	0.51	32.57
						1350	382	0.54	0.41	32.93	237	0.05	0.27	33.11
	Soccer	477	13.98	0.76	28.09	750	472	12.73	0.73	28.76	441	11.90	0.76	29.09
						1350	416	10.36	0.68	27.96	322	5.37	0.69	31.21
	Tempete	536	19.28	0.82	25.02	750	523	18.56	0.80	23.91	469	16.05	0.82	26.30
						1350	453	14.49	0.71	23.40	342	10.09	0.71	27.64

a function of the considered aggregation strategy, of the maximum RTP packet size constraint and of the characteristics of the video sequence (see Table 1). The Foreman sequence, for example, is made of small-size NALUs, hence many NALUs can be encapsulated into one RTP packet. Thus, the number of accesses to the channel is noticeably reduced even if the maximum RTP packet size is set to half the MTU (750 bytes). The opposite considerations hold for sequences with large NALUs such as Football: multiple NALUs can be fitted into a single RTP packet only in a few cases and only if the threshold on the maximum RTP packet size is set close to the MTU. Obviously, the *Optimal* strategy is more effective than the *TID-based* strategy in reducing the number of channel accesses, which drops from 522 to 237 in the case of the Foreman sequence.

Simulations show that the byte loss rate and the packet triptime decrease when the number of accesses to the channel is reduced, both in error-free and noisy channel conditions. For instance, the Coastguard sequence shows 8% BLR reduction with the *TID-based* strategy and 9% with the *Optimal* strategy in error-free channel conditions. Such BLR reductions are due to the increased number of correctly received Class D packets, that is the lowest priority traffic class which encompasses mainly enhancement NALUs with TID equal to four. When the number of packets offered to the network decreases, the transmission queues become shorter and therefore less packets are dropped due to timeout expiration or queue saturation. Shorter queues imply lower triptimes, as in the

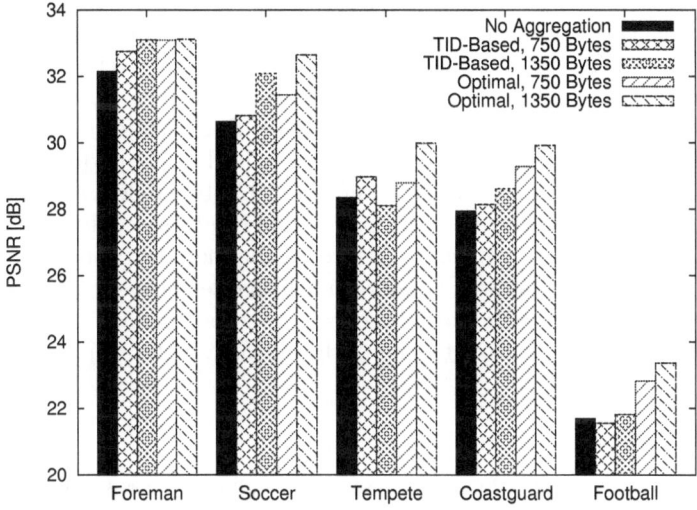

Fig. 5. PSNR of the received videos as a function of the NALU aggregation strategy and the limit to the size of the aggregated RTP packet

case of the Foreman sequence, whose average triptime drops from more than half a second to less than 0.2 s in the case of the *Optimal* strategy.

Figure 5 shows that the *TID-based* and the *Optimal* strategies achieve PSNR gains in excess of 1 dB with respect to the standard case where no NALU aggregation is performed. Clearly, the *Optimal* strategy achieves the most considerable improvements in video quality (1.6 dB with the Football sequence and 1.8 dB with Tempete). The performance of the *TID-based* strategy is in many cases close to that of the *Optimal* strategy, whose computational complexity is however higher. The increase in video quality is mainly due to the correct reception of a higher number of NALUs with TID equal to four. Such NALUs noticeably increase the temporal resolution of the video sequences characterized by high levels of picture motion such as Soccer or Football, which in fact boast the most significant PSNR increases.

Aggregating NALUs improves the communication quality even in noisy channel conditions: a single transmission error in a packet causes the whole packet to be discarded, so the larger the packet is, the more likely it is affected by errors. However, if the aggregation strategy is effective in reducing the number of transmitted packets, the packet loss rate increase due to channel errors can be counterbalanced by the 802.11 automatic retransmission mechanism, leading to a reduction of the application-level BLR and an increase in video quality.

6   Conclusions

This work aims at improving the quality of H.264/SVC video communications over 802.11e networks. In particular, this work addresses the inefficiencies of the 802.11 channel access mechanism when a high number of small packets are offered to the network, as it is often the case of SVC communications. Two strategies with different computational complexity and effectiveness are proposed to aggregate multiple SVC data units prior to their transmission, also taking into account their perceptual importance. Both strategies produce RTP packets which are compliant with the RFC standards. The performance of the proposed strategies is evaluated in a domestic 802.11e wireless ad hoc network scenario in various channel conditions. Simulation results show that the proposed low-complexity strategy significantly improves channel utilization and provides significant quality gains as well as more timely packet delivery with respect to a traditional SVC communication. Moreover, the performance of the proposed low-complexity strategy is close to that of the optimal, high-complexity, strategy.

References

1. IEEE 802.11e Standard, 2005: Wireless LAN Medium Access Control and Physical Layer specifications Amendment 8: Medium Access Control Quality of Service Enhancements
2. Wiethölter, S., Emmelmann, M., Hoene, C., Wolisz, A.: TKN EDCA Model for ns-2. Telecommunication Networks Group, Technische Universität Berlin, TKN-06-003 (2006)
3. Wenger, S., Wang, Y.K., Schierl, T.: RTP Payload Format for SVC Video. IETF draft-ietf-avt-rtp-svc-15 (work in progress) (November 2008)
4. UCB / LBNL / VINT: Network Simulator - NS - Version 2.31, http://www.isi.edu/nsnam/ns
5. ITU-T & ISO/IEC Standard: Advanced video coding for generic audiovisual services. Annex G, Scalable Video Coding H.264 & 14496-10 (November 2007)
6. Ksentini, A., Naimi, M., Guéroui, A.: Toward an improvement of H.264 Video transmission over IEEE 802.11e through a cross-layer architecture. IEEE Communications Society Magazine 44, 107–114 (2006)
7. Schwarz, H., Marpe, D., Wiegand, T.: Overview of the Scalable Video Coding Extension of the H.264/AVC Standard. IEEE Transactions on Circuits and Systems for Video Technology 17(9) (September 2007)
8. Fiandrotti, A., Gallucci, D., Masala, E., Magli, E.: Traffic Prioritization of H.264/SVC Video over 802.11e Ad Hoc Wireless Networks. In: Proceedings of the 17th International Conference on Computer Communications and Networks, ICCCN 2008 (August 2008)
9. ISO/IEC 8802-11, ANSI/IEEE Std 802.11 Standard: Wireless LAN Medium Access Control (MAC) and Physical Layer (PHY) Specifications (1999)
10. Niculescu, D., Ganguly, S., Kim, K., Izmailov, R.: Performance of VoIP in a 802.11 Wireless Mesh Network. In: Proceedings of the IEEE INFOCOM (April 2006)
11. Bianchi, G.: Performance analysis of the IEEE 802.11 distributed coordination function. IEEE Journal on Selected Areas in Communications 18(3), 535–547 (2000)
12. H.264/AVC Reference Software JM 11.0, http://iphome.hhi.de/suehring/tml/doc/lenc/html/index.html
13. H.264/SVC Reference Software JSVM 9.4, http://ip.hhi.de/imagecom_G1/savce/downloads/SVC-Reference-Software.htm

Framework for Personal TV

André Claro[1], Paulo Rogério Pereira[2], and Luís Miguel Campos[3]

[1] Instituto Superior Técnico, Taguspark Campus, Av. Prof. Dr. Cavaco Silva,
2744-016 Porto Salvo, Portugal
[2] Inesc-ID, Instituto Superior Técnico, Technical University of Lisbon, Rua Alves Redol, 9.
1000-029 Lisboa, Portugal
[3] P.D.M.&F.C, Av. Conde Valbom n. 30, Piso 3, 1050-068 Lisboa, Portugal
andre.claro@ist.utl.pt, prbp@inesc.pt, luis.campos@pdmfc.com

Abstract. This paper proposes a study of the IPTV world, focusing on network and system architectures, video codecs, network protocols, services and quality assurance. Based on this study, a new framework for Personal TV was developed. This new system is designed primarily to provide new personalized services to the user. In the architecture of this framework there are three main elements: Clients, Aggregators and Producers. The most important element is the aggregator that provides all video contents from the producers to its clients. The client has the possibility of creating his own channel that is sent to the network through its aggregator, of creating customized channels that can be viewed by other clients, among other features. The architecture designed and developed is based on new and studied concepts. It was tested to prove its viability, to assess its performance and to draw conclusions about its scalability, based on functional tests, compatibility tests and performance tests.

Keywords: IPTV, IP, Television, Services, Personalization, Availability.

1 Introduction

Telecommunications companies are investing in ways to offer its customers triple-play services and some of them are starting to think about integration with mobile services. This means that there is a convergence at the network level for IP (Internet Protocol) to offer the different services. In an all-IP network, personalized and interactive services can be provided very easily. IPTV (Internet Protocol Television) is the technology responsible for this convergence and this technology will bring a variety of new business opportunities.

Based on the growing number of users of IPTV and also of VoIP (Voice over IP), a growing migration of existing technologies to IP technology is expected. Associated to this convergence, new services and new business opportunities may arise. IPTV emerges as the future of fixed or mobile television. Many of the carriers around the world are investing millions in technology in order to increase profit from their networks and to compete with rivals over providing a wider range of services. There are other companies which invest in Internet TV, which is based on the same concepts of IPTV, but with a best effort service.

R. Valadas and P. Salvador (Eds.): FITraMEn 2008, LNCS 5464, pp. 211–230, 2009.
© Springer-Verlag Berlin Heidelberg 2009

The main goal of this paper is to investigate and study the protocols and concepts involved in IPTV with the aim of widening horizons and designing a Framework for Personal TV, e.g. personalized television services, where the user is a key element in the architecture.

The user can create its own personal channel and personalized channels. Channels generated by users can be distributed through the network as any other channel. The architecture is composed of three main elements: Clients, Aggregators and Producers. The aggregators are interconnected in a network of aggregators, with fault-tolerance and load-balancing, that distributes the TV contents for all clients. This architecture can operate on a private network using unicast or multicast, or on the Internet, where bandwidth is more limited and only unicast is available.

The developed architecture was tested, and results were analyzed to draw conclusions about the performance and functionality of the system.

2 State of the Art

The telecommunications companies are evolving their networks to all-IP networks. IPTV is one of the concepts and technologies used for this convergence. "IPTV" is the secure delivery of high-quality multichannel television, on-demand video and related multimedia contents, via a dedicated network, to a consumer electronics device such as a set-top box, a computer, or a portable device that is served by broadband IP access. This is in contrast to Internet TV (or Internet Video) – be it live TV or video-on-demand – brought to users via the open Internet on a best-effort basis, where multicast is not available [1][2][3].

2.1 Network and System Architectures

The ITU-T, through its study group on IPTV, presents a functional architecture based on existing technologies and concepts or on NGN (Next Generation Network). Fig. 1 shows the four functional domains [4].

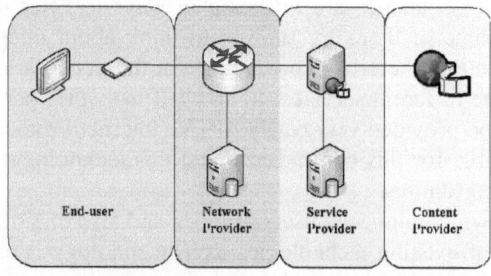

Fig. 1. Functional domains of IPTV according to ITU-T

In an alternative model proposed by the BSF (Broadband Services Forum), an IPTV system is based on the following elements: Video Head-end, Network Core, Network Access and the Home Network [5].

Content Distribution Networks (CDN) are a good way for IPTV operators to spread their services offer outside their networks and consequently expand the list of channels and services available. CDNs use "multicast" on the application layer, aiming independence from the network infrastructure. The architecture Prism (*Portal Infrastructure Streaming Media*) shown in Fig. 2 is an example of a CDN [6].

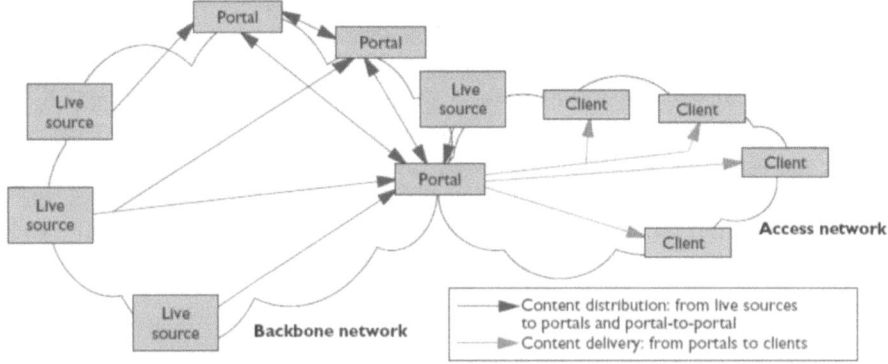

Fig. 2. Portal Infrastructure Streaming Media architecture

Peer-to-peer IPTV networks are another kind of architecture, more suitable to the public Internet. In peer-to-peer networks, client nodes are also servers so as to distribute the server load over all nodes in the network.

2.2 Video Codecs

Video encoding is a major factor required to offer IPTV, given the constraints at network level. H.264 is the video compression standard that is being adopted in new solutions for IPTV. It tends to replace MPEG-2 standard that is still widely used in some systems. VC-1 is also a valid alternative and competitor to H.264, by presenting a similar compression and image quality, amongst other advantages [7].

2.3 Network Protocols and IPTV Services

The transport of encoded video/audio in IP networks can be made based on several technologies. The main ones are: RTP (Real Time Protocol) [8] and MPEG-2 Transport Stream (TS) [9]. The use of native RTP has several advantages over MPEG-2 TS: bandwidth preservation, flexible audio/video stream selection, better resistance to errors, improved quality of service feedback through the use of RTCP and improved services [10].

IGMP is a multicast protocol used by IPTV, which allows minimizing packet replication in the network when distributing the same contents to a group of destinations. Generally, IGMP version 2 or 3 is used [11][12].

Other protocols used by IPTV are: RTSP (Real Time Streaming Protocol) [13], a signaling protocol for VoD services; SAP (Session Announcement Protocol) [14], a multicast session announcement protocol; SIP (Session Initiation Protocol) [15], a signaling protocol used to control sessions between users; SDP (Session Description Protocol) [16], a protocol for describing the session characteristics, used in other protocols such as SAP, RTSP, etc.

Based on these protocols, the following services may be offered for IPTV [17]: Linear TV (audio, video and data), Linear TV with Trick Modes (pause, replay), Multi-View service, Time-shift TV, Pay Per View (PPV), Video/TV on Demand (VoD), Download Based Video Content Distribution Services (Push VOD), Content download service, Personal Video Recording (PVR) service (network or client-based), Interactive TV (iTV), Linear Broadcast Audio, Music on Demand (MoD) including Audio book, Learning (education for children, elementary, middle and high school students, languages and estate, etc.), Information (news, weather, traffic and advertisement etc.), Commerce (security, banking, stock, shopping, auction and ordered delivery, etc.), Telecommunication (e-mail, instant messaging, SMS, channel chatting, VoIP, Web, multiple), Entertainment (photo album, games, karaoke and blog, etc.).

2.4 Quality Assurance Service

In order to ensure the services described above, it is necessary to have the best architecture, the best video codecs and protocols for efficient data transport over the network.

There are many components to guarantee the quality of service (QoS). For the management of QoS, the following components are important: management of subscribers and content, management of the capacity of the network and systems, service level monitoring, image quality metrics, network monitoring and testing.

The concepts of QoS and quality of experience (QoE) are also important. QoE is a more complex concept compared to QoS which includes rates, delays, losses, latencies, etc. QoE includes the total end-to-end system effects and is measured based on user perception of the quality of the image. Fig. 3 shows the QoS and QoE domains [18].

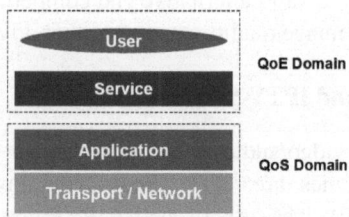

Fig. 3. QoS/QoE Domains

3 Proposed Architecture

3.1 System Description

The designed solution is based on an end-to-end architecture, with the aim of providing new IPTV services. This system relies on network aggregators responsible for the distribution of audiovisual content to clients. This IPTV platform enables the management of multiple video sources by an operator. This architecture requires the existence of three entities, as shown in Fig. 4: Aggregators, Clients and Producers. It was built on the vision of ITU-T and of Content Distribution Networks (CDN).

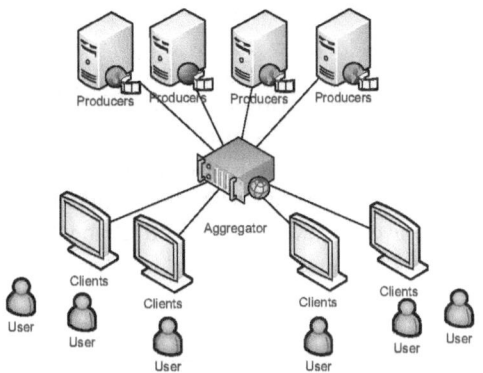

Fig. 4. Simple Architecture

The aggregator is the most important element of the system, which in addition to distributing the content to other aggregators, offers the following new IPTV services to the client: Personal Channels and Personalized Channels. Personal Channels are composed of media generated by the user from a webcam or from a video file selection. Personalized Channels are channels produced by combining the channels available in the producers.

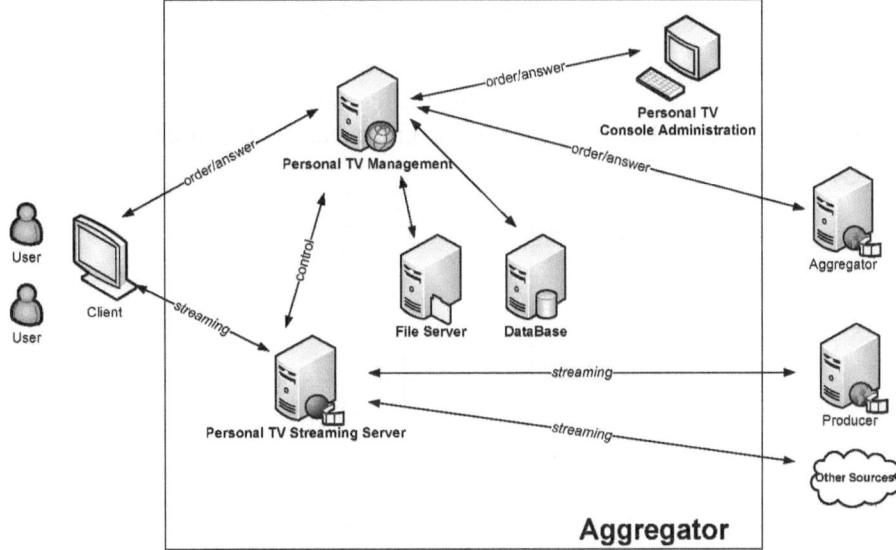

Fig. 5. Aggregator Details

3.2   Aggregator Details

The Aggregator is composed by the main components shown in Fig. 5: Personal TV Management, Personal TV Streaming Server, Database, and Personal TV Administration Console. The figure also shows the types of interactions between the different components.

Personal TV Management

The Personal TV Management, which will be called "PTvM", is the main component of the aggregator, as shown in Fig. 6. It is responsible for all of its management and control. The main features are: communication with others aggregators; control of the streaming Server (PTvSS); communication with clients; management database; monitor applications; create/remove/change personalized channels; add/remove/ change producers; communication with billing systems and in interfacing with other systems.

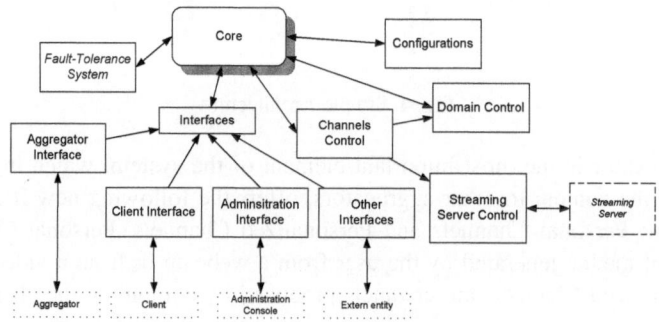

Fig. 6. PTvM Class Diagram

Data Model

The domain shown in Fig. 7 represents the identities that are stored in a persistent database. A simple data model was drawn for the features developed.

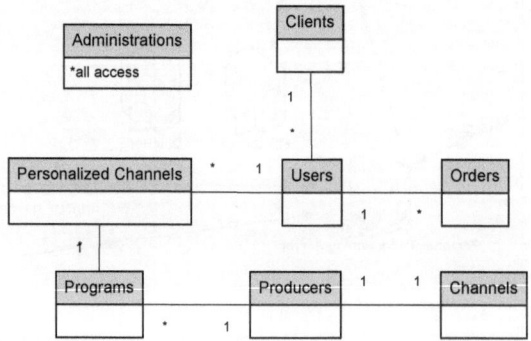

Fig. 7. PTvM Domain

High Availability

To solve the problem of aggregator redundancy and load balancing, it was decided to introduce a High Availability solution in the architecture. In each geographical zone there are several Aggregators available. Instead of clients connecting directly to an Aggregator, they connect to a Load-Balancing Server, which forwards the requests to the Aggregator that is online and has higher available capacity.

Streaming Server

The Personal TV Streaming Server (PTvSS) is the server for content streams, which sends and receives content, from clients, producers and other aggregators. This element will support multiple protocols for transportation and various encodings, in order to achieve greater interoperability with other systems. It should also allow the transmission/reception in unicast and multicast.

3.3 Client

The Client is a simple entity, with a user interface and a video player, allowing the user to use all the functions provided by the aggregator to which it is connected. The Video Player allows the customer to send video streams to the network, working as a produced in order to offer the Personal TV service. The video streams generated may come from a webcam or from a selection of stored videos. Fig. 8 represents the client's class diagram.

3.4 Global Vision

This global system is shown in Fig. 9. It has aggregators as main entities, forming a network responsible for content distribution between producers and clients. Clients connect to an area of aggregators, through a load-balancing system that distributes clients through the aggregators of that area.

Producers only send streams of video to the aggregators. Producers get the flow of video over any IP network, with or without of quality of service guarantees, or via satellite (DVB-S), or generate the flow from a file or DVD. The producer is a simple streaming server.

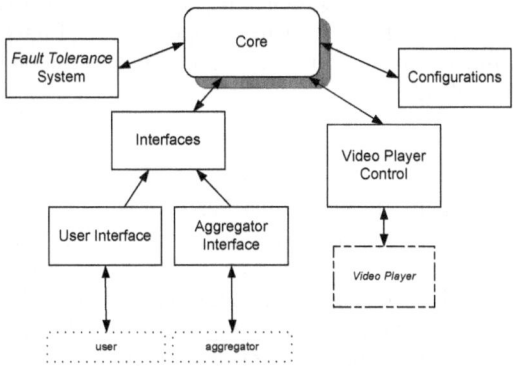

Fig. 8. Client Class Diagram

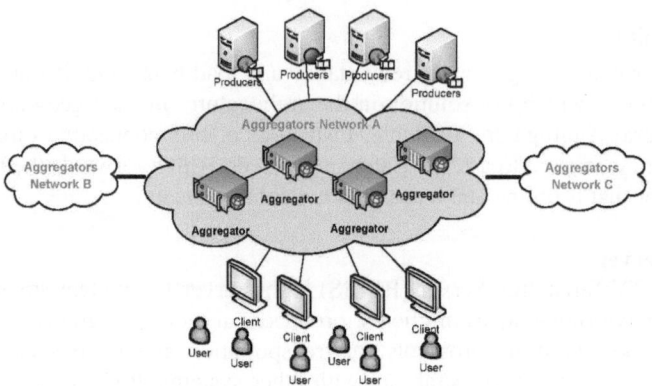

Fig. 9. Global IPTV Architecture

4 Application Analysis

The following streaming servers were analyzed: Adobe Flash Media Streaming Server 3 [19], Darwin Streaming Server [20], Feng [21], Helix DNA Server [22], Live 555 Media Server [23], MPEG4IP [24], VLC [25] and Windows Media Server [26].

Based on the analysis carried out on those applications, it was decided to use VLC as a server for streaming and also as a client in the designed architecture referred in the previous section. This choice was due to several factors, including: the flexibility of being able to make changes as the application evolves; the lack of licensing costs; compatibility with various platforms namely Linux, a free operating system. Another aspect that was considered was the number of updates and fixes that the development team regularly does, allowing constant improvements and increased functionality of the solutions implemented around them. This program has a free software license based on General Public License (GPL).

5 Implementation

The prototype developed is based on free applications. The majority of applications are based on GPL, which greatly facilitates its use and does not require any license. The development was carried out with the support of the Linux operating system, Ubuntu. It is a multi-platform solution and it is designed to run on other systems such as Windows and Mac OS X.

5.1 Aggregator

In the Aggregator, the main element is the Personal TV Management, which is responsible for the management and control of the entire IPTV system. This element controls the streaming server, the database and a file server. There is also a management console (Personal TV Console Administration), which enables remote management and administration of the aggregator. Fig. 10 shows the aggregator illustrating the connections between the various elements and the protocols used for each one.

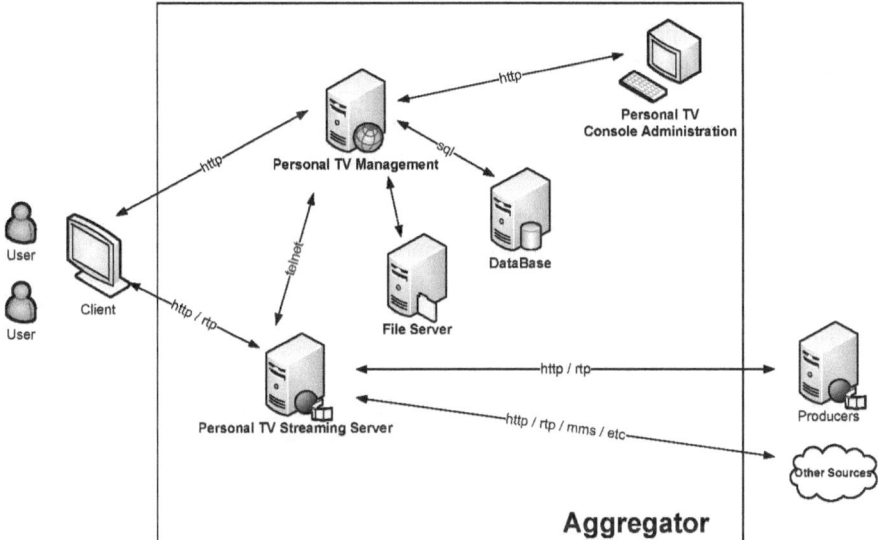

Fig. 10. Aggregator Details

Personal TV Management (PTvM)

This Application Server manages everything that happens in the aggregator.

The following applications and libraries were used to develop the PTvM: Java (programming language), MySQL (database management), JWSDP [28] (Java Web Services Developer Pack, web services library), Hibernate [28] (database connection library), Apache Commons Net [29] (telnet library).

High Availability

The High Availability solution is implemented based on the LVS (Linux Virtual Server) [31] and the HAProxy [30] proxy, as shown in Fig. 11. Based on the LVS, the following applications were used: keepalived [28] and Heartbeat [29].

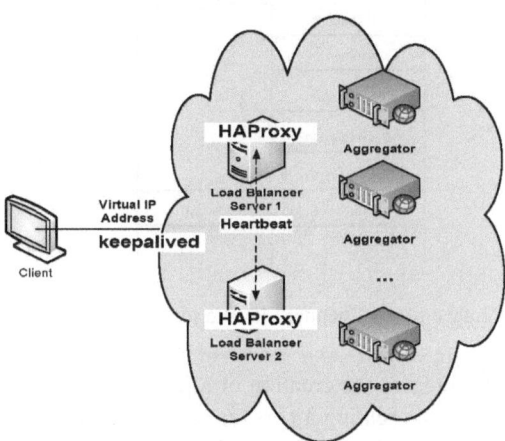

Fig. 11. High Availability Solution

The keepalived application is responsible for the setup of virtual addresses. The Heartbeat application is responsible for verifying the status of the other load-balancing server. In case of failure of the primary server (master), the secondary server (slave) takes the lead, and sets the virtual address, through keepalived. The HAProxy redirects the requests for aggregators that are in operation and with lower loads.

If there is any problem with the aggregator to which the client is connected, the load-balancing server forwards requests to another server that is available. This process is transparent to the Client.

5.2   Client

The client was also developed based on the Java programming language and the VLC video player. A user interface based on the command line was developed. The JWSDP library to program web-services communication was used to assure the communication with the aggregators. There is a fault-tolerance system for quick recovery in case of an error on the server. For personal channels, many configuration parameters can be changed, for example: codec, rate and send mode (unicast or multicast).

5.3   Communication Diagrams

In this section, some of the most important concepts of the implementation of communication between components of the system are presented. There are several diagrams representing the communication between the components of the architecture: Client, Aggregator (PTvM and Streaming Server).

The communication between a Client and its Aggregator, to request the channel list and to select a channel to be viewed, is represented in Fig. 12.

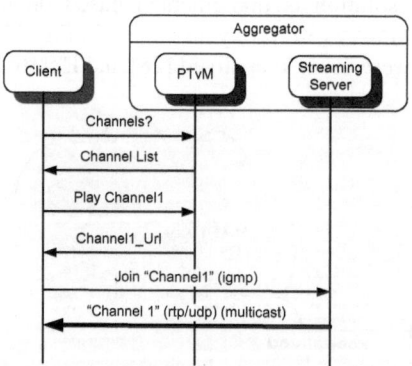

Fig. 12. Request of the Channel List and joining a Multicast Channel

Fig. 13 and Fig. 14 represent the creation of a personal channel and a personalized channel, respectively. When creating a personal channel, the client sends a request to the aggregator and it asks the client for the video stream. In the personalized channel creation, the client sends new channel information and programs of the new channel to the aggregator. In both cases, after the channel configuration on the Aggregator, the channel information is propagated to the other aggregators.

6   Tests and Evaluation

Tests were designed to demonstrate the main features of the implemented solution, and assess system performance. The tests are divided into three types: functional tests, compatibility tests and performance tests.

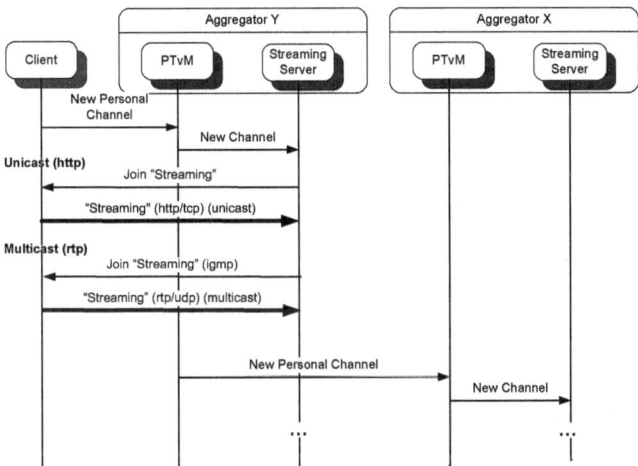

Fig. 13. New Personal Channel

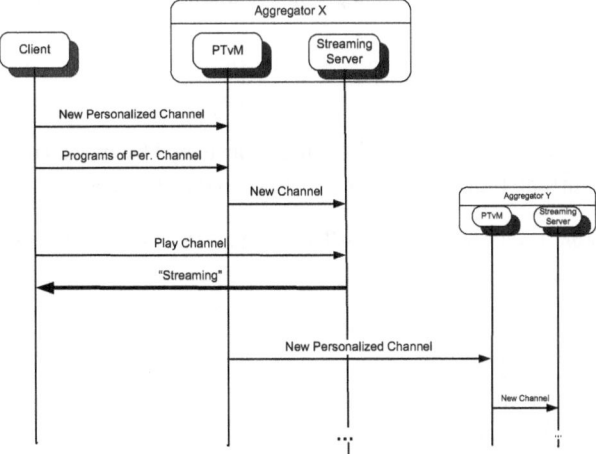

Fig. 14. New Personalized Channel

The collection of data from tests was done with Cacti [34] (software for monitoring computer networks and systems), TOP (Linux application) and a software module developed for the PTvM.

Tests were conducted in a laboratory using twelve computers, connected to an Ethernet switch (Cisco Catalyst 2950) at 100 Mbps. The computers used had a 3 GHz processor with 1 GB of RAM (Random Access Memory) and Ubuntu Linux operating system. They were prepared to serve as clients and aggregators.

6.1 Functional Tests

First, the features provided by Aggregators to Clients were tested: view a channel, create a personal channel, create a personalized channel, etc. These tests were based on a simple architecture, with a client and an aggregator that is linked to several producers, as shown in Fig. 15.

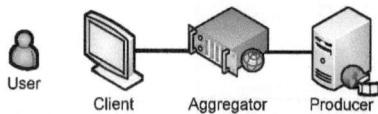

Fig. 15. Functional Tests Scenario

Based on the tests carried out, all the features have been well implemented, both on the client side and on the server side. In the section on performance tests, the performance of each function will be analyzed.

6.2 Compatibility Tests

All implemented features are compatible with Linux, Windows and Mac OS X operating systems, and only some parameter changes in streaming server execution are needed.

6.3 Performance Tests

Function Execution Time
The performance of the implemented features was tested by measuring their execution times. No concurrent function execution was tested. The results shown in Table 1

Table 1. Function Execution Times without Competition (seconds)

Function	Average	Standard Deviation
Open/Change Channel	1.030	0.177
Create Personal Channel	4.660	0.178
Create Personalized Channel	3.210	0.202
Create Personalized Channel by Category	4.210	0.228

indicate that the time to open/change a channel takes on average 1 second. A personal channel takes on average 4.6 seconds to be available to be viewed by another client. For a personalized channel by category, the creation delay is between 4.1 and 4.35 seconds, with 95% confidence.

Client Performance

The performance of the client in terms of CPU processing, memory used and network speeds was tested. The Client starts by receiving a video stream (at 2 Mbps) and after sends a video stream (at 1 Mbps) for the network (Personal Channel). Fig. 16, Fig. 17 and Table 2 show the test results.

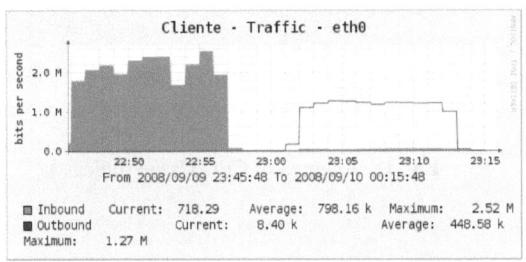

Fig. 16. Client Network Traffic

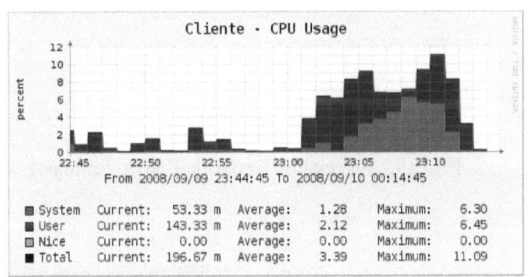

Fig. 17. Client CPU Usage

Table 2. Client Application Used Memory (MBytes)

	Average	Standard Deviation
Receive Channel	20.19	1.53
Send Channel	22.08	0.71

It can be observed that the client requires a minimum computational load when it is receiving or sending a stream. Using a computer or embedded system with fewer capabilities, the developed client solution would continue to function properly without affecting the quality of the user's viewing experience. Even with increased resolution (2 or 4 Mbps), the increase of processing is not relevant.

Aggregator Performance

The performance of the Aggregator was tested with different clients in two parts: unicast streams and multicast streams to the clients. Fig. 18 to Fig. 21 show the test results. The aggregator sends a 2 Mbps video stream encoded with H.264. The memory used by the aggregator applications is presented in Table 3.

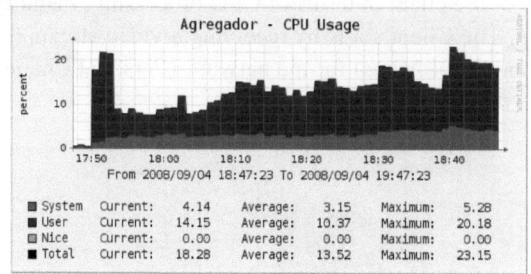

Fig. 18. Aggregator CPU Usage (unicast)

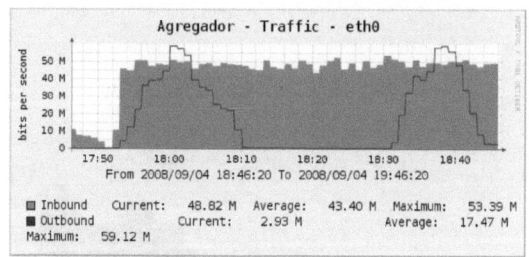

Fig. 19. Aggregator Network Traffic (unicast)

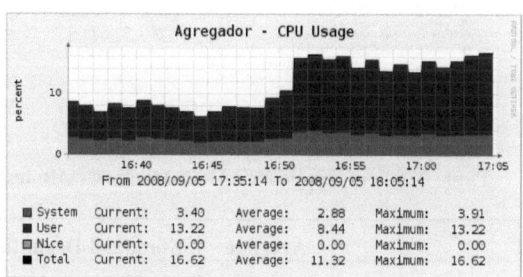

Fig. 20. Aggregator CPU Usage (multicast)

Based on the results, it is seen that when the aggregator sends unicast streams, it needs more average processing than while sending multicast. The average speed of incoming traffic is identical in both tests because the Aggregator is connected to the same number of producers.

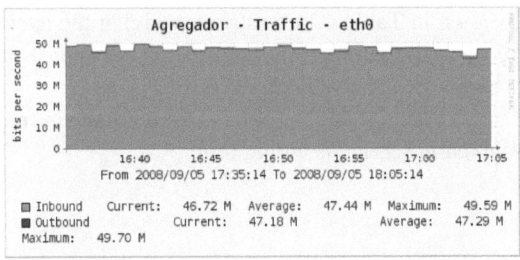

Fig. 21. Aggregator Network Traffic (multicast)

Table 3. Aggregator memory usage (MBytes)

	Clients	Average	Standard Deviation
Multicast	10	163.210	1.901
	20	167.029	1.131
Unicast	10	272.540	4.219
	20	274.695	4.658

The processing increases with the number of producers linked to the customer as each corresponds to a channel stream being received. When the solution is working on multicast, less memory is used compared to the unicast case. This probably happens because there is a single socket for multicast, with a single thread, while in unicast there is a socket and a thread for each client.

Aggregator Communication
The communication time between Aggregators was measured. For this test, the personal channel functionality was used and the time necessary for a channel to be available on another aggregator was measured.

Based on the results, one aggregator takes on average 0.6 seconds to receive information of a new aggregator channel and to provide that channel to clients. With ten aggregators in series, the last aggregator takes on average 4.93 seconds with a variance of 0.012 seconds to receive and provide the channel contents to its clients. With aggregators in parallel, the time to propagate a new channel increases slightly with the number of aggregators connected and the increase is mainly due to the increase in CPU processing of the aggregators that send the information.

Failure Recovery
Client recovery in response to aggregator failure was also tested. In case of failure of an aggregator, its clients should quickly connect to another aggregator in the same zone and continue receiving the video streams. First, a network with only two aggregators, in which clients were all connected to the same aggregator was tested. Afterwards, a network with 6 aggregators was tested. Both topologies are shown in Fig. 22.

The results are shown in Table 4. It was observed that the average recovery time is about 1.5 seconds, when the number of client varies between 1 and 10, and the number of aggregators available is between 1 and 5.

The results are satisfactory, if such problems do not happen frequently, because 1.5 seconds without television service is a long time.

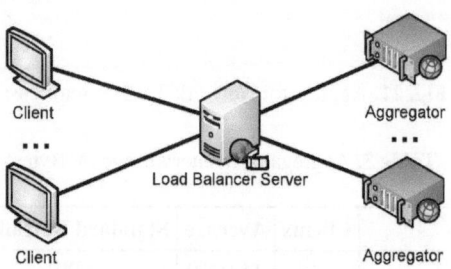

Fig. 22. Failure Recovery Test Scenario

Table 4. Failure Recovery Times (seconds)

Aggregators at start (after)	Clients	Average	Standard Deviation
2 (1)	1	1.37	0.183
	10	1.81	0.3
6 (5)	1	1.32	0.132
	10	1.48	0.148

Scalability of the Solution

The system CPU processing at the aggregator depends more on the number of producers than on the number of clients connected. Fig. 23 shows an estimate of CPU processing versus the number of producers. The increase is linear.

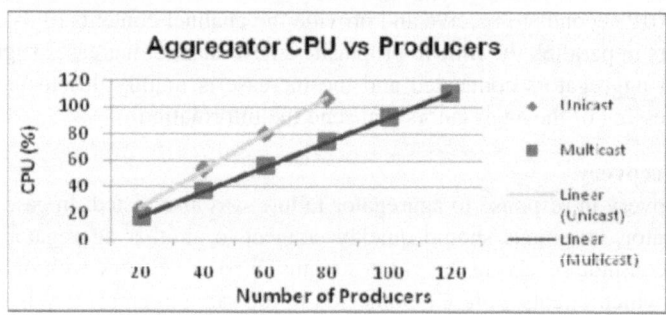

Fig. 23. Estimate of CPU vs Producers

From Fig. 23 it can be observed that the maximum number of producers connected in unicast may be just over 75 per aggregator for the type of computer used (3 GHz). In multicast, that number is higher: around 110 producers.

The aggregator memory necessary is mainly related to the number of clients. Thus we can conclude that in multicast the memory occupation is lower than for unicast. Fig. 24 shows that the memory used grows approximately exponentially with the number of clients.

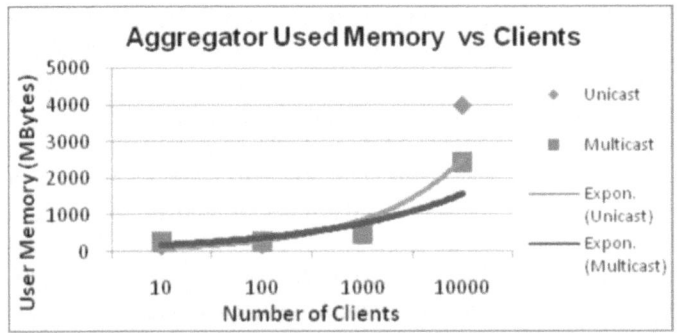

Fig. 24. Estimate of Used Memory vs Clients

It can be concluded that the network scalability depends on the method used by the aggregator to send the video streams to the clients. If the method is unicast, the number of users is limited by the maximum speed of the network. With multicast, the number of channels transmitted depends on the number of channels in the aggregator and does not depend on the number of clients.

Tests Conclusions

The number of clients supported by an aggregator depends on the server hardware where the aggregator is operating. But the network's bandwidth also limits the scalability. The use of multicast can solve part of this problem if the number of channels is not very high.

In order to distribute the CPU processing for the developed solution, the following steps may be done: put PTvM and the Streaming Server on different computers and also increase the number of computers for each of these applications (both redundancy and performance of the system are increased).

7 Conclusion

This paper presents a study of the concepts and technologies used in IPTV: architectures, network protocols, video codecs and others concepts. The IPTV area is having a large expansion and development. A Framework for Personal TV was developed, based on an architecture that assumes the existence of three entities: Clients, Aggregators and Producers. This architecture is designed to be scalable, fault-tolerant and provide customized audiovisual content and new services to clients.

A prototype system for IPTV was designed, implemented and tested, with the aim of providing the client the following features: display of real-time channels, creating custom channels and personal channels. This opens a door to new services implementations. The prototype is designed based on the study of IPTV architectures, especially the view of ITU-T and CDN.

Much of the system implementation was performed in the aggregator, which has the following features: serving clients with the best quality and performance, communication with other aggregators to share information and video content, user statistics and others features. The developed solution can run on a private network with guaranteed quality of service or on the Internet. In this case, it is always limited by the speed of the network and of its connection to the user. The absence of guarantees of quality of service can also restrict the flows and impose limitations.

Based on the prototype testing, it is concluded that the system responded as expected. With the machines used (CPU at 3GHz), customers get times in the order of 1 second to open or change channels and the times of recovery observed in a client in the event of an aggregator failure are between 1 and 2 seconds. To create a personalized or personal channel, a time between 3 and 5 seconds is needed until the channel becomes available. It was found that an aggregator can support 75 producers (each corresponding to an IPTV channel) in unicast, or 110 producers in multicast, because the processing and memory required in unicast is higher than in multicast. As multicast is not supported by all networks, unicast should be used for sending video streams. This method makes the scalability of the solution, and the number of users more limited. Client applications consume at most only 12% of CPU time (at 3GHz) and 25 MB of RAM. The time for sharing information between aggregators was found to be less than 1 second for a new channel to be available in neighboring aggregators. The solution can scale to any number of channels by the use of a network of aggregators by zone. The network of aggregators can be connected to other area zones to share resources between them.

Future Work

As future works, a list of activities expected to strengthen the idealized concepts and the features of the system is presented. There is a need to integrate the solution developed with a cache system. Other features needed are: graphical interface for the client and a secure architecture to support the framework developed by using secure communication channels for communication between the elements of the system.

The solution can also be integrated with other systems, such as: DRM / CA (Digital Right Management / Conditional Access System), AAA (Authentication, Authorization and Accounting), BSS (Business Support System), CRM (Customer Relationship Management), Order Management (management applications), Network Inventory/Service Activation. Integration with other systems can be the base for offering new services such as: VoIP, video conferencing, customized advertising to each user and others services.

The possible evolution of a solution based on open source software for a professional solution based on a commercial streaming server could also be a topic for future work. Naturally, the concept of the designed architecture should be kept: Clients, Aggregator and Producer.

IPTV should be considered seriously because with this concept and technology, television will evolve more over the next five years than it evolved in the last twenty years.

References

1. Cooper, W., Lovelace, G.: IPTV Guide - Delivering audio and video over broadband. Informitv (2006)
2. IPTV-news.com. IPTV News, The Magazine for all your ipTV information, 1st edn. (November 2007), http://www.iptv-news.com/
3. Good, R.: IPTV vs. Internet Television: Key Differences (April 4, 2005), http://www.masternewmedia.org/
4. Focus Group. FG IPTV-DOC-0115 - Working Document: IPTV Architecture. ITU-T (July 31, 2007)
5. BSF. IPTV Explained - Part 1 in a BSF Series, http://www.broadbandservicesforum.org/
6. Cranor, C., Green, M., Kalmanek, C.: Enhanced Streaming Service in a Content Distribution Network. IEEE Internet Computing, 66–75 (July 2001)
7. Sunna, P.: AVC/H.264 - An Advanced Video Coding System for SD and HD broadcasting. EBU Technical (2005)
8. Schulzrinne, H., Casner, S.: RFC 3550 - RTP: A Transport Protocol for Real-Time Applications. IETF (July 2003)
9. DVB-ETSI. ETSI TS 102 034: Transport of MPEG 2 Transport Stream (TS) Based DVB Services over IP Based Networks. ETSI (March 2007)
10. ISMA. Planning the Future of IPTV with ISMA, http://www.isma.tv/technology/white-papers/ ISMA-IPTV_whitepaper_V11_2006-09-14.pdf
11. Juniper Networks. Introduction to IGMP for IPTV Networks (October 2007), http://www.juniper.net/solutions/literature/white_papers/ 200188.pdf
12. Fenner, W.: RFC 2236 - Internet Group Management Protocol, Version 2. IETF (November 1997)
13. Schulzrinne, H., Rao, A.: RFC 2326 - Real Time Streaming Protocol (RTSP). IETF (April 1998)
14. Handley, M., Perkins, C.: RFC 2974 - Session Announcement Protocol. IETF (October 2000)
15. Rosenberg, J., Schulzrinne, H.: RFC 3261 - SIP: Session Initiation Protocol. IETF (June 2002)
16. Handley, M., Jacobson, V.: RFC 4566 - SDP: Session Description Protocol. IETF (July 2006)
17. IPTV Focus Group. FG IPTV-DOC-0116 Working Document: IPTV Service Scenarios. ITU-T (July 2007)
18. Kishigami, J.: The Role of QoE on IPTV Services. In: 9th IEEE International Symposium on Multimedia (2007)
19. Flash Media Streaming Server. Adobe, http://www.adobe.com/products/flashmediastreaming/

20. Darwin. Darwin Streaming Server, http://dss.macosforge.org/
21. Feng. The RTSP/RTP streaming server,
    http://www.lscube.org/projects/feng
22. Hellix. Hellix Community, https://helixcommunity.org/
23. Live555. LIVE555 Media Server, http://www.live555.com/
24. MPEG4IP. MPEG4IP: Open Source, Open Standards, Open Streaming,
    http://www.mpeg4ip.net/
25. VLC. VideoLanClient, http://www.videolan.org/vlc/
26. Windows Media Server. Microsoft WMS 9,
    http://www.microsoft.com/windows/windowsmedia/forpros/
    server/server.aspx
27. Java Web Services Developer Pack,
    http://java.sun.com/webservices/downloads/previous/
28. Hibernate, http://www.hibernate.org/
29. Apache Commons Net, http://commons.apache.org/net/
30. LVS. Linux Virtual Server, http://www.linuxvirtualserver.org/
31. HAProxy. HAProxy - The Reliable, High Performance TCP/HTTP Load Balancer,
    http://haproxy.1wt.eu/
32. KeepAlived, http://www.keepalived.org/
33. Heartbeat, http://linux-ha.org/
34. Cacti. Cacti - The Complete RRDTool-based Graphing Solution,
    http://www.cacti.net/

Author Index